WALDEN ON WHEELS

WALDEN ON WHEELS

On the Open Road
from Debt to Freedom

KEN ILGUNAS

New Harvest • *Houghton Mifflin Harcourt*
BOSTON   NEW YORK   2013

This edition published by special arrangement with Amazon Publishing

For information about permission to reproduce selections from this book, write to Permissions, Houghton Mifflin Harcourt Publishing Company, 215 Park Avenue South, New York, New York 10003.

www.hmhbooks.com

Library of Congress Cataloging-in-Publication Data is available.
ISBN 978-0-544-02883-8

Printed in the United States of America
DOC 10 9 8 7 6 5 4 3 2 1

For Mom and Dad

Ten percent of all author royalties will be donated to The Wilderness Society, an advocate and protector of America's wild places.

Remember who you wanted to be.

— BUMPER STICKER ON VOLVO IN
THE HOME DEPOT PARKING LOT

NOTE FROM THE AUTHOR

....

I tried to write this book as true to life as possible. It bears mentioning, though, that I changed the names and altered details of several characters. And, on a few occasions, I took liberties with the chronology when I did not think the authenticity of the story would be compromised.

End of 2009 Spring Semester
Duke University
DEBT: $0

ON A SPRING EVENING in 2009, in a campus parking lot at Duke University, I was lying on the floor of my van, trying to hide from view.

This is it, I thought. *They know.*

I was wearing nothing but my boxer shorts, splayed out on the carpet, face up, like a scarecrow toppled by the wind. My arms and legs took up pretty much my whole central living space: the section of the van in between the bed in the very back and the driver and passenger seats up front. Next to my head was my three-drawer plastic storage container, which, in addition to serving as my counter for cooking, held pretty much everything I owned: stacks of cereal boxes, bags of rice and beans, and a jumble of miscellaneous items, including sewing supplies, utensils, duct tape, and recycled Ziploc bags.

It was a sunny day. As on all sunny days, it was far hotter inside the van than it was outside. It was unreasonably hot. Scalding hot. Africa hot. The van was a moist 98°F womb, and I, a red-faced fetus wishing for an orifice to wriggle out of. Between the North Carolina heat and the very realistic possibility of *they* knowing about my secret, my glands thought it was a good idea to discharge all reserves of sweat, darkening the hue of my underwear, imprinting a moist outline of my body onto the carpet, and encasing me in a slick film of perspiration.

Moments before, when I'd caught a glimpse of the vehicle that had parked next to me, I'd been cooking my evening meal on an isobutane backpacking stove on top of my storage container. As soon as I saw his car through my window, I pulled the blinds down, turned off the stove, and flattened myself on the floor. It was a security guard's patrol car.

I wondered, *What was it that gave me away? Did someone see me get into the van? Has someone ratted me out?*

Please don't knock on my door. Please! Don't knock!

I held my breath as the security guard shut off the engine. His door croaked open and slammed. His heels clicked like Gestapo jackboots across the asphalt.

If the security guard was inspecting the van, as I thought he was, then I knew things must have looked awfully strange to him. All the window blinds had been pulled down. There was a large black sheet of fabric hanging behind the passenger and driver's seats that prevented him and all other passersby from seeing what was in the back of the van. Though he couldn't see me because of the blinds and the sheet, he could freely inspect the van's front area, where he was sure to make note of the laundry basket on the passenger seat, spewing out pant legs and sweaty white tees. Or he might pause to wonder why the windshield had fogged up (from the steam from my cooking pot). Or why odors of broccoli and onion were leaking out of the windows I left ajar.

I knew that, any second, he'd tap on my window and I'd have no choice but to answer his questions and explain to him why I was half-naked, why I was cooking a meal in a van, and why I'd been secretly living in a campus parking lot for the past four months.

I knew how the dominoes would fall: the university's administration would revoke my parking permit, I'd be banned from all campus parking lots, and, without the van, I'd be forced to adopt some conventional and unaffordable style of living – one where I'd have to rent an apartment, hang decorative curtains, and buy a rug to tie the room together. And I'd have to break the promise I told myself I wouldn't break four months earlier: I'd have to go back into debt.

For the whole 2009 spring semester, as a graduate student in Duke's liberal studies program, I'd been secretly living in my van. And this was the moment I'd been dreading.

INTRODUCTION,

OR

WHAT'S A "VANDWELLER"?

A "VANDWELLER" — AS DEFINED by vandwellers – is, essentially, a person who lives in a vehicle. The type of vehicle the vandweller lives in doesn't matter, so long as the lifestyle is simple and the home has wheels.

It may come as a surprise to some readers that there are more than a couple of people living in their vehicles right now. At this moment, one of them might be reading Edward Abbey with the glow of her headlamp in a Walmart parking lot. A man and his dog may be driving west on the I-90 in a 1985 Dodge cargo van with the windows rolled down. A few retirees might have formed a wagon circle in the desert outside of Quartzsite, Arizona.

They're parked inconspicuously in front of hotels and mechanic shops. They're asleep on your city street one night and gone the next. They're America's modern-day vagabond; the twenty-first century's tramp and hobo, drifter and gypsy. They're people who take pride in living lives of thrift, adventure, and independence. They're unburdened by belongings, unfastened from earthly foundations, and unruffled with the prospect of going to the bathroom in an empty Gatorade bottle. They're vandwellers.

Over the years, I've learned that – behind a vandweller's decision to move into what is often a cramped, smelly, heatless, air-conditioning-less, "you mean it doesn't come with an espresso wine tower?!" vehicle – there is always a story. And because the vandweller made such a decision because they were lured to the road by powerful, romantic longings, or because he or she was forced to resort to desperate measures in desperate times (or because "there's just something wrong with him," as my mother would put it), it's usually a good story.

A great many people have asked me what it's like to live in a van. They've inquired how I fed myself, how I stayed warm, and how I kept it a secret, among other practical queries. But the most frequent question asked pertains to *why* I did it.

From other vandwellers' stories, I've learned that the vandweller doesn't become a vandweller simply by purchasing a van. Rather, some personal change or transformation must first occur. The answer to the question about why I lived in a van is this book, which means that the following story isn't so much about a van but about student debt, and wilderness, and all the people and places and journeys that have made me the person I am today, who happens to be okay with tight quarters and dubious hygiene.

So in order to properly tell my story, I ought not begin with the moment I bought a Ford Econoline in a used car lot outside of Raleigh, North Carolina, in January 2009. Rather, I should start years before – four years before, to be exact – when I was an indebted twenty-one-year-old about to finish my fourth year of college.

Part I

....

DEBTOR,

or
My Attempt to Pay Off $32,000
in Student Debt with a Useless
Liberal Arts Degree

1

················

CART-PUSHER

April 2005–University at Buffalo
DEBT: $27,000 AND GROWING

I DREAMED OF THE GRIZZLY BEAR. It was my only recurring dream. Ever since I'd turned sixteen, I would dream this dream over and over again. It was always the same: A half mile south of my parents' home, in a neighboring suburban development, I'd happen upon a grizzly bear grazing on someone's lawn. It would spring up onto its hindquarters, inspecting me from the top of his bulky blond tower of fat and fur. I'd look back at it, paralyzed, awestruck, exhilarated.

That was it. I had this dream repeatedly. And afterward — when I'd be lying in bed in that half-dreaming, half-awake state — the dream would feel so real that I'd often wonder if it was in fact a dream, or if it was a distant memory that I could only vaguely recollect. I'd always wanted to believe that I'd really seen the bear, but I knew that that was impossible because: 1) There are no grizzlies in the suburbs of western New York, or anywhere near New York for that matter; and 2) I'd somehow gone the first twenty-one years of my life without experiencing anything even remotely interesting.

It was my fourth year of college. Many weekday evenings and

weekend mornings, I'd tie an orange apron around my waist and collect orange shopping carts strewn across a giant Home Depot parking lot in Niagara Falls, New York. I'd gather a dozen at a time, press them together, pivot them around curbs, and march them to the vestibule inside. When all the carts had been accounted for, I'd work inside the store, stacking lumber, folding cardboard, reorganizing shelves, emptying garbage bins, and lending a hand to any customers who needed help loading drywall or bags of Quikrete. I was a cart-pusher.

For your ordinary college student, pushing carts wasn't the worst job local industry had to offer. I'd considered it a step above jiggling a WE BUY GOLD sign for the local pawnshop and a few steps below the indentured servitude of an unpaid internship, where students, though unpaid, could at least hope that their career paths were leading them to a more prosperous destination than stacking four-by-fours in the lumber department.

I spent upward of thirty hours a week at the Home Depot, making $8.25 an hour. I was certainly more frugal with my paycheck than your average student, yet these were my profligate years, when I wasted a good chunk of my hard-earned money on a daily Dr Pepper, the occasional CD or DVD or video game, or – if I had the weekend off – long road trips to get drunk with friends at distant colleges. Mostly, though, my money was used for responsible purposes, like paying the various bills needed to keep my car running and the occasional $100 here, $100 there "offering" to my already-massive and still-growing $27,000 student debt.

I was able to keep the car running, but what little money I was able to put toward my debt always felt negligible – pointless even. It was like throwing a glass of water on a burning building. It was a sacrifice to appease the gods, but a pitiful, emaciated, bony goat of a sacrifice. Such paltry offerings, I worried, might seem less a declaration of submission – which it was – and more an affront to the debt's greatness, which just might make it angrier, prodding it to swell with interest.

There was no controlling my debt. It grew and grew and

grew. It was a mountain of coins that rose with interest every month to such staggering Himalayan heights that it made me feel — when I thought of its immensity — small and weak and insignificant. It was huge. My debt was a black hole, a swirling abyss that sucked from my clutches all my hopes and dollars and dreams.

My debt wasn't as bad as other students' debts, but because I was soon going to enter the real world with an unmarketable degree (a B.A. in history and English) and because I had absolutely no idea how I was going to pay it off, the debt, to me, was more than a mere dollar amount. It was a life sentence. And soon enough, I'd be behind the bars of the great American debtors' prison, alongside the other 36 million Americans or so who'd similarly sentenced themselves to decades of student debt.

I was worried about letting the debt get any bigger, so I pushed carts and pushed carts some more. I worked full time during winter and spring breaks, as well as on weekends. When I got home I would — inside a hoodie powdered with Quikrete and stained with paint — hurriedly leaf through textbooks and hastily type up research papers.

While I'd balanced school and work reasonably well in previous years, the lifestyle had begun to take its toll during my fourth year of college. I'd grown tired of spending a huge portion of my week at a place I hated. I tired of reciting the "Home Depot chant" at obligatory monthly store meetings. I tired of the bottom-of-the-food-chain position I had, which gave head cashiers the liberty to assign to me some of the more unpleasant tasks required to keep a big chain store humming, like removing dead pigeons from the lumber section, mopping up overflowing toilet water, and sweeping the remains of torn bags of concrete whose particles would dry out my eyeballs and coat my nose hairs with a pale gray pollen. More than anything, I tired of the winter holiday season, which, if memory serves me right, begins a little after Labor Day at the Home Depot. Upon listening to Gloria Estefan sing "The Christmas Song" for the third time in an hour, my mind would be consumed with morbid fantasies. I'd imagine myself derailing the toy train that chugged above the cash registers by whipping a hammer at it, or, better yet, hanging myself with an electrical cord from the rafters out of protest, if just to shame the suits in corporate into changing store Christmas music policy, thereby granting me the solace of knowing, in my dying moments, that I'd performed at least one useful service for mankind.

Between commuting to school, the long hours at work, the papers, and the exams, I had little time for study and hardly any for sleep. Like many college students, I began to decompose into a paler, flabbier, oilier, much more caffeinated version of myself. My eyes turned bloodshot, new wrinkles webbed across my face like creases in a catcher's mitt, and my hair began to fall out. When I lay in bed reading, I'd obsessively pluck out what few chest hairs I had like some mistreated parrot. At some point, I'd picked up a minor case of Tourette's syndrome, and when I thought no one was listening, curse words would dribble from my lips. In class, I had to fight the inexplicable urge to jam the point of my pen into the back of my hand.

I'd always considered myself "well adjusted," so this whole falling apart thing was new to me. And the extent of my deterioration was especially made apparent on a morning in late April during finals week, when something rather unexpected and unbelievable and potentially life altering occurred.

I heard a voice.

At the time, because I didn't yet have the luxury of hindsight, I'd failed to realize that my physical and psychological deterioration was due in large part to a decision I'd made years before.

It all began in August 2001, when I decided to participate in one of the great annual migrations known to man: alongside millions of fellow eighteen-year-old Americans, I had graduated from high school and was going to college.

My high school class and I moved like a school of fish: we

graduates were capable of going off on our own, in whatever direction we chose, but something demanded we all swim as one, curving, cutting, sashaying together, wiggling our way to college. Except for a few miscreants, we all ended up in college.

In high school, if someone asked me what my "plans" were, I'd click into brainwashed robot mode: my body would become rigid, my pupils would dilate, and in a monotone, I'd recite, "I-will-go-to-the-best-college-I-can-get-into. No-matter-the-cost." At some point, I'd convinced myself that going to college was what I really wanted to do. So my best friend Josh and I migrated to Alfred University, a pricey private college in southern New York.

Josh had graduated from high school with high honors, which qualified him for a large financial aid package that reduced the cost of his tuition. My performance as a student, however, could euphemistically be described as "unremarkable." I was ranked seventy-seventh of my two-hundred-student high school class and was probably regarded by my teachers to be just a notch above "slacker" – only slightly more capable than the students who were funneled into afternoon vocational programs so they could get a head start on their manual trade educations. I didn't do clubs, didn't do volunteering, didn't do student government, didn't do music. Apart from playing on the hockey and football teams, I didn't do much of anything. I drifted through the weary waters of high school on a dinghy of disdain.

But after graduating from high school and learning how much I would have to pay for just my first year of college, I thought for the first time that maybe I should have spent the last four years of high school doing something more productive than spending my nights playing video games and masturbating till three in the morning.

My first year at Alfred would cost me $18,450.

It didn't occur to me to think about how strange it was that the government, my college, and a large bank were letting me, an eighteen-year-old kid – one who didn't know what "inter-

est" was (or how to work the stove for that matter) – take out a
gigantic five-digit loan that might substantially alter the course
of my life.

Taking out student loans was a momentous event in my life,
yet I don't have the faintest recollection of the event. I know
it happened because I definitely went into debt, but I don't at
all remember signing any forms, shaking any hands with financial aid officers, or noting the frown that was surely fixed on my
mom's face as she cosigned the loans with me – which was, by
the way, probably a daunting prospect to her, as I'd given her no
indication that I'd one day exhibit traits of industry, ambition,
or responsibility.

I do, however, remember *not* hearing any warnings about the
consequences of debt or the likelihood of a bleak postgradu-
ation job market. And I do remember hearing, from a chorus
of voices, that "student debt is good debt" and that "money
shouldn't stop you from going to the school you want to go to."
Like everybody else, I listened.

I never actually thought about why I was going to college,
or why I was about to take out thousands of dollars in loans
for it. Like most eighteen-year-olds, I cared little for books, or
higher learning, or anything that had to do with school. I was
told that school was for "developing yourself" and "preparing
for a career." Why would I want to do either of those?

I hated school. We all hated school. Why were we all so willing to go back? For Josh and me, it probably had something to do
with our image of what college life was like. We reveled in ludicrous fantasies of enjoying passionate and fleeting escapades
with the opposite sex. We tacked a few posters of scantily clad
(though hardly scandalous) women on our dorm walls, hoping
the presence of their glistening paper bods would somehow
draw real (hopefully glistening) women into our beds. Alas, our
dorm room never quite became the laboratory for sexual exper-
imentation we'd dreamed of. We liked to blame our failure on
our embarrassing dorm decor, which matched because our
mothers, regrettably, did our school shopping for us with each

other, but the truth was, while color-coordinated lamp shades, rugs, and sheets certainly did not communicate prowess to the fairer sex, we failed to lure any girls into our room because we were both painfully shy, awkward, and boring.

After that first year of college, in a rare moment of sagacity, I realized that Alfred was costing me way too much and that I wasn't the type of person who'd be making the big bucks someday. So I decided to transfer to the University at Buffalo (UB), where I could get a part-time job and save money by commuting from my parents' home.

I still had no better idea why I was in school or why I had just spent almost $20,000 on one year of college. I told myself that incurring student debt was like puberty or a midlife crisis: it was an unavoidable nuisance, a ticket required for admission to the next stage of adulthood, a burden I had to clumsily carry up the socioeconomic ladder. But as I continued to take out more loans to pay for tuition, books, and my car, I slowly began to grasp that dealing with this debt was going to be more than some paltry inconvenience.

For a moment—because I began to wonder why I'd enrolled in school—I considered dropping out. But only for a moment. I knew there was no way I'd be able to pay off my debt with the sort of money I made pushing carts. If I was going to flounder in a sea of red ink, I hoped my degree, at least, would be a plank of driftwood with which I could keep my head above the surface.

So I found myself, more or less, trapped in school.

The University at Buffalo was a relatively affordable state school that cost me about $7,000 a year in tuition—most of which could be paid off with the wages I earned pushing carts during the school year and landscaping full time during summers. At UB, my sophomore and junior years passed much the same way as my first. I fulfilled my general education requirements, sampled courses from different departments, and attended lectures in large auditoriums with hundreds of fellow students. Occasionally, I would enjoy a course or a lecture, but for the most part I coasted through college much the same way I coasted through high school. But then, at the tail end of my junior year, something strange happened, and it happened without warning.

I started to give a shit. I started to enjoy school.

In the English department, I read Shakespeare; in the history department, I studied the Constitution and the Founding Fathers. During my senior year, my classes got smaller. My classmates and I had thoughtful, often passionate, discussions. I wrote for the university's newspaper, befriended a couple of professors, and took an unpaid summer internship in Virginia interviewing D-day veterans and another one, later on, writing stories for Buffalo's alternative weekly newspaper. I put everything I had into every essay. I promised myself that I'd read every page of assigned reading. At night, after work, I'd go on the computer and type essays. My mother, who'd bring me up a plate of food for dinner, would take note of my puffy, sleepless eyes and say, "It'll be over soon, Ken. You'll graduate in just a year."

"I know," I'd say, "but I love school."

It was a renaissance. Every day I could feel my "horizons expanding," as they say. I was writing and speaking more clearly. I began having new ideas and asking new questions. College, for the first time, felt like a place where I belonged.

College was helping me shed my high school slacker skin, revealing someone who had passions and ideas, convictions and dreams. But while it was freeing some part of me, it was also chaining my ankles to the steel balls of debt, which I knew I'd have to drag through the halls of Career World for the foreseeable future.

And oh, what a bleak future it was going to be! How would I pay for it all? How could I afford a car, rent, a cell phone, health insurance, gas, electricity, Internet access, three magazine subscriptions, a gym membership, and a movie a month, not to mention my student loan payments???

But it wasn't just the carts, or the exams, or the debt that left me feeling battered and frayed and a little crazy; it was that I began to see that I lived in a free country but couldn't say I knew what it felt like to feel "free." And while I owned plenty of stuff – a car, DVDs, CDs, clothes – I never felt like I owned my own life. College had helped me see how everything, for my whole life, had either been predetermined or planned out: I went to high school because I was forced to; I went to college because I was supposed to; and now I'd enter Career World because I was financially obligated to.

Yet who was I to complain about anything? My adolescence was an American idyll. Not once did I have to deal with an exploding volcano, an ethnic cleansing, or a potato famine. I was never molested, bullied, or forced to stick my tongue to a frozen pole. Nor was I ever obliged to hunt down my biological mother or to screech, pubescently, "You're not my father!" My problems were, in comparison to the rest of the world's, privileged, first-world problems – problems I was lucky to have. Yet, despite my good fortune, there was something missing, some desire unsatisfied, some glaring need that my normal suburban upbringing failed to fulfill.

I spent much of my boyhood playing video games and watching epic adventure movies for hours on end – movies like *Braveheart*, in which mud-spattered warriors got to do manly, gallant things. While delivering the *Buffalo News* as a paperboy, I'd imagine myself cleaving off the arm of an enemy; telling a woman, "You and no other"; and screaming, "Freedom!!!" before being disemboweled for some righteous cause. Like almost any boy, I wished to live in a world where there was real adventure, real glory, and real sacrifice – just as it was on-screen.

My mom was a nurse and my father was a factory worker who took the night shift. He put in, typically, ten hours of overtime a week. Every day, after work, they'd come home and watch TV. They did the same thing day after day, week after week, year after year. My mom would watch *Oprah* and *Judge Judy*, and my father would watch *Coronation Street* – a British drama series that aired on the local Canadian station. My parents were comfortably domestic, bearing few desires to travel, try new things, or take adventurous detours off old, rutted paths.

I grew up thinking it was normal for a married couple to never sit on the same couch, hug each other, or demonstrate even the vaguest expression of intimacy. Because I wanted proof that things like romance and passion and desire existed in real life, I may have been the only child in history to have actually wanted to see his parents doing it.

Before every Christmas, I'd ask my mom to buy me a claymore sword—the sort that Mel Gibson carried in *Braveheart*. After years of patiently waiting, I came downstairs one December morning to see a long rectangular package. I unwrapped it excitedly with full knowledge of what was inside. The sword, though, lost some of its luster when I learned from the receipt my mom had left in the box that she hadn't hunted down the fabled claymore that once crossed the Ilgunas family shield, but that she'd bought a cheap Pakistani version for $30 on eBay. Still, I had my sword, and when no one was home, I'd carry it into the backyard and swing it around behind our aboveground swimming pool. It's always adorable to watch a little boy play make-believe, but it gets a touch desperate and disturbing when he's eighteen.

Now, after four years of college and a good deal of personal growth, I was twenty-one but still living in my parents' home, still bedding under the same revolving Super Mario fan that had whirred me to sleep as a six-year-old, and still pushing carts. I'd never done a drug, broken a law, or diverted from the path prescribed to me by social and parental expectation. I'd hardly left home, except for my internship in Virginia and a very rushed road trip to California the summer before. And while I yearned for new experiences, I recognized that every year I was getting more and more into debt and becoming less and less free.

I became afflicted with a burning restlessness that stirred up irrational, impractical dreams and coaxed out strange, subconscious voices. At home, I'd slap the globe on the computer desk and skim fingertips over spinning topographies. At the campus library, I'd wander over to the atlas shelves, always ending up on the page with the map of Alaska. I'd picture myself driving up the Alcan Highway across northern Canada and west to Alaska. I'd be driving on a gravel road that meandered around pristine mountain lakes, endless spruce forests, and snow-topped mountains. I wanted to stand on one of these peaks and take in the frozen white sea of undulating summits and know – if just for a fleeting second, maybe upon viewing a herd of caribou, or gazing, teary-eyed, at the northern lights – what it felt like to feel free.

I wanted to drive to Alaska more than anything in the goddamned world. And getting there, as far as I was concerned, was my life's purpose, my dream of dreams, my ultimate adventure.

And every spring I told myself that this was the summer I was really going to do it. Yet it always seemed to make more sense to spend the summer pushing carts to pay tuition or to work at unpaid internships to fill out my résumé. It made sense to blaze a path toward a secure, stable, comfortable life. It made sense . . .

Alaska didn't make *any* sense. I had no idea why I wanted to go so badly. I knew nothing about the state. Yet I was drawn to it as if by some unbending law of physics, lured with the same intensity of passion I felt for the fairer sex, beckoned as if it were a pair of moonlit thighs. Alaska pulled me by my shirt collar north toward a land far different from the suburbs I'd grown up in.

On my commute to college, I'd sometimes fantasize about driving past my school, hopping on the thruway, and heading north. I wouldn't look back, and I wouldn't stop until I'd escaped the sprawling suburbs, the car dealerships, the parking lots, and the starless, smoggy night skies. I'd leave behind my family and friends, the papers, the orange apron – my stale suburban life.

• • •

It was one of the last days of the 2005 spring semester. I drove my red 1996 Oldsmobile Cutlass Ciera to college as I did every morning. (Because I was one of thousands of students who commuted to UB, I had to arrive an hour before class in order to wait for a parking spot to open up.) With everyone still in class, the campus seemed like a set suitable for a postapocalyptic zombie movie. It was a bleak, lifeless, white-lined landscape where the cars were abandoned in the eerily quiet daylight hours. Apart from the swarm of gray storm clouds inching across the sky like ghost-driven blimps, and a plastic shopping bag that performed parabolas in the wind, the campus was deathly still. In the background, on my stereo, was Canadian alt-rock musician Matthew Good playing his dreamy "Near Fantastica."

This is when I heard the voice.

It was a whisper: raspy yet distinct, quiet yet audible – a voice that I could clearly identify as my own, except that I hadn't spoken aloud. It shared with me four simple words – each just one syllable long – the first of which was my name, *"Ken,"* followed by a three-word message.

I swiveled around to the backseat to see if anyone was in my car. No one. Frantically, I opened the door and pressed the side of my face against the asphalt to locate the feet of the unseen speaker.

There wasn't a soul in the parking lot.

With gravel stuck to my cheek, I pulled myself back in the car, looked in the rearview mirror, and saw, looking back at me, a young man with a pale face and a purple bag under each eye. I looked pitiful, wearing a shamefaced "I just got my ass kicked in front of my friends" expression. *Look at what you've become,* I thought. I was a loan drone: existing, yet hardly living.

A *voice?* In *my* head?

I felt like I was losing it. I'd never heard a voice before. I'd never had any "spiritual" or "paranormal" experiences. I hadn't been drinking, and I'd never done any drugs, so I certainly wasn't on anything.

It seemed like the voice was issuing a command or a directive. But the message was so vague. What did those three words mean?

I sat there, scared, panicky, and vaguely diarrheic, watching my fellow students pour out of the buildings and make their ways toward their cars.

A week later, the semester ended and it was time to go back to work for the summer. I'd been promoted to the position of delivery coordinator, which would be a high-stress, high-responsibility, high-labor job that, curiously, came with no raise.

I put on my standard Home Depot outfit – a pair of jeans and a green polo shirt – got in my car, put the key into the ignition, and buckled my seat belt. But I just sat there staring at the wheel. I sat there as I had a week before in the UB parking lot. I couldn't bring myself to turn the ignition. I felt like I was about to enter into some inhospitable environment. Going into the Home Depot would be like walking into a burning building, or swimming to the bottom of a sea, or freezing solid, floating aimlessly through outer space. Thrashing my forearm against the rim of the wheel as hard as I could, I screamed. It was an angry, throaty, "my child was just eaten by lions" scream.

This was my last summer before I had to begin paying off my debt. If I spent it working at the Home Depot, I'd be declining some rare gift. I'd be a kid at fat camp turning down a Milky Way trafficked in by a softhearted counselor. I'd be a prisoner spending his yearly conjugal visit session alone in his cell. I'd be a bum tossing the winning lottery ticket into a flaming barrel.

I felt like a character in a story, but a sorry, cowardly, timid, unassertive character. Shouldn't this be a turning point? Shouldn't this be the moment when I get my shit together? The moment when I change and grow? When I get the hell out of here? If I didn't go now, would I ever?

I thought of those three words and felt a terrible sense of urgency sweep over me, like the encroaching shadow of a UFO that darkens the landscape over which it passes. Things in my parents' driveway were serene. I was parked next to a basketball hoop propped up on a crisply mown lawn. The trees were a lush spring green and the sky a healthy cloud-pocked blue. Yet I felt the presence of this dark premonition – one that would give me one chance, one moment, one "now or never" opportunity to either give in to my insanity or forever accept my role as loan drone and cart boy. *Time is running out,* I felt anxiously. Tomorrow, I'd have a real job. Tomorrow, I'd be forty. Tomorrow, I'd be dead.

I'd once heard that we are nothing but our stories. Forget the blood and bones and genes and cells. They're not what we are. We are, rather, our stories. We are an accumulation of experiences that we have fashioned into our own grand, sweeping narrative. We are the events and people and places to which we've assigned symbolic meaning. And it's when we step outside our stories that we feel most lost. If we take the wrong path at the classic fork in the road – and fail to act in a literary sort of way – our story falls apart. Words run off the page. Paragraphs are cluttered with red markups. Pages fall out of the binding. And we lose a grip on our identity.

Despite outward appearances, I knew I wasn't just a loan drone, or a cart-pusher, or some beer-drinking college student. So while it might have seemed expected of me to drive to the Home Depot that day and accept my station, along with the dull comforts of the status quo, it would also feel, strangely, out of character.

So I got out of the car, went back into my house, and took a pair of scissors and cut up my orange apron. And then I began packing a suitcase full of clothes, books, and camping gear.

A week later, I turned out of my parents' driveway and drove down the road. I passed my friends' homes, the Home Depot, and the sprawling suburbs of western New York. As I came to UB – that mammoth campus that had both freed and shackled me – I decided that this time, I'd keep going. This time, I wouldn't look back.

CHEECHAKO

May 2005–Coldfoot, Alaska

DEBT: $27,500 AND GROWING

AFTER I QUIT THE HOME DEPOT, I made a list of every Alaskan lodge, campground, and tourist operation that I could find on the Internet and called each to find out if any of them were still hiring for the summer. But no one was. I was making these calls in May and all the business owners I spoke with said they'd done their summer hiring months earlier. One lady, though, who sounded like she was still looking for workers, asked me what my "trade" was.

It was a question that hit me like a hockey puck to the back of the knee—the one body part my equipment didn't cover—because it reminded me of the one thing I'd neglected to prepare for throughout my college education, which was learning an actual skill. Her polite rejection to my honest answer ("I'm afraid I don't have a skill"), I hoped, wasn't an augury of things to come.

Unmarketable or not, I was determined to get up to Alaska that summer because I'd figured it might be the last summer in my life to do something adventurous. I'd be graduating from college next year, and to pay my debt, I knew I'd have to soon stagger around in office hallways and conference rooms as one of the lobotomized undead, a "young professional."

Near the bottom of my list of Alaskan work sites was a camp in a town called Coldfoot (population 35), whose website proudly advertised its status as the "Farthest North Truck Stop." It was in the middle of the Brooks Range, 60 miles north of the Arctic Circle and 250 miles from the nearest stoplight in Fairbanks. Along with servicing the truckers who haul equipment up the Dalton Highway to the Prudhoe Bay oil fields, the truck stop also runs a small tourism operation.

During the summer months, the Slate Creek Inn, a one-story, fifty-two-room motel, was often full of tourists who traveled on Holland America and Princess Cruises bus lines up from Anchorage. Even though their trips were several weeks long and even though they'd spent nights in innumerable hotels and lodges along the way, Coldfoot still left such an impression on them that several tourists felt inspired to leave pithy comments on hotel review websites.

"This motel was nasty," one guest said.

"One step better than a tent."

"Truly dreadful," bemoaned a third.

Because Coldfoot's remote location made it difficult for the camp manager to attract seasonal employees, he was still seeking a pair of lodge cleaners. I interviewed, got the job, leaped in the air jubilantly, and asked my freshman year roommate, Josh, if he wanted to join me. Josh, though, balked at the last minute, so I persuaded my other good friend Paul to take a leave of absence at his job at a UPS distribution center, promising the adventure of a lifetime.

I was going to Alaska.

Because my car wasn't suited for a long cross-continent road trip, I asked my father if I could borrow his SUV. He agreed. I lent him my car, and to make sure I felt some semblance of freedom and self-reliance (as any self-respecting adventurer ought to feel), I promised that I'd pay his monthly car payments for the whole summer.

It took us seven days and 4,500 miles to get from New York to Coldfoot. The memory of our road trip is bordered with a dreamy blur, perhaps because we traveled in an almost chronic state of sleeplessness. It was more or less a twenty-four-hour-a-day operation: while one slept, the other drove. We lived on peanuts, beef jerky, and Dr. Bob – a cheaper supermarket brand of Dr Pepper. We sang Tom Cochrane's "Life Is a Highway" in the Alberta prairie, came to a halt in front of a herd of bison in the Yukon, and, upon running out of conversation topics, had long discussions about which animal we'd most want to be and, later, "do."

If just for the summer, we thought of ourselves not as poor students but daring adventurers. And when we crossed the Alaskan border and arrived in Coldfoot, all we wanted was to keep going and make our grand entrance into the wild.

But first came a week of cleaning. For eight hours a day we made beds, scrubbed showers, and brushed toilets for $8 an hour. On the summer solstice (June 21), which is more or less an observed holiday in the arctic, we cleaners got together with the rest of the Coldfoot crew – the guides, maintenance workers, cooks, servers, dishwashers – and all got drunk on Miller High Life on the gravel bank of the Koyukuk River. As the midnight sun carouseled around us, I tottered around a bonfire and screamed, at the top of my lungs, "I'M!!! IN!!! ALASKA!!!"

Paul and I were eager to test our "manhood" in the wild, so we decided to climb the biggest mountain within a reasonable distance of the truck stop. Scanning a giant topographic map of the region at a nearby ranger station, we confidently placed our fingers on Blue Cloud, a 5,910-foot peak just inside the Gates of the Arctic National Park and Preserve, which would be an off-trail, backcountry hike ten miles each way.

Aside from the stroll we took up a small hill when we drove through Yellowstone National Park, this would quite literally be the first hike of our lives. Paul and I had grown up in western New York—a topographically challenged plain of flatland, on top of which sat a horrid landscape of subdivisions, chemical plants, and abandoned warehouses. Our suburban neighborhood was called "Country Meadows," which was mostly rural when we were kids but had been, as we grew up, bulldozed and paved over to make way for new houses and tracts. What adventures we could embark on had to take place in virtual video game worlds on television screens or, at best, within the borders of football fields and hockey rinks. So no opportunity ever called for us to filter water, light a fire, use a compass, or do anything even remotely "manly."

Paul was a closet metrosexual who wore pink polos (because he got them "on sale") and insisted—to anyone who inquired—that his arms were, in fact, naturally hairless. I was woefully inept at just about everything, indoors and out. When Paul and I tried to set up our tent on a pre-trip trial run, the finished product looked like a crumpled tissue. We were "cheechakos"—a term the Alaskan natives used during the gold rush that meant, more or less, "the idiots from down south who don't know what they're doing." More precisely, we were suburbanites, only familiar with landscapes of strip malls and sublots, soccer fields and sprawl.

I hoisted onto my shoulders my brand-new backpack, filled with a large camcorder, an aging three-person tent that a friend had lent to us, a cumbersome bear-resistant food container (holding a few peanut butter and jelly sandwiches), and a sleeping bag that I sloppily strapped to the pack's exterior. The ranger station didn't have extra topographic maps, so they gave us a faded photocopy of the area on a regular-sized sheet of paper.

Paul and I, already struggling with the heft of our packs, walked along an old Caterpillar mining trail that petered out after a mile, and then ducked and elbowed our way through a dense thicket of alder trees. Upon reaching a clearing, we looked at each other excitedly. Sizing up the landscape that spread

out before us, we could see there weren't any streets or buildings. We couldn't hear any hum of traffic or planes screeching
through the sky. We'd finally escaped the babble of man and his
machine. We'd finally entered the wild.

We stood in the middle of a wide, rolling green valley. The
country was grassy and open except for a procession of gnarled
spruce and wizened birch that followed a creek. Perhaps tired
of the arctic's fleeting summers and bitterly cold winters, the
trees seemed to have embarked on a forlorn exodus out of the
valley in search of balmier pastures. Farther back, there were
mountains — glorious mountains with seams of shining snow
branching down from flint-gray peaks. It was as if we'd walked
into a black-framed motivational poster with something like
PERSEVERENCE printed at the bottom.

The Brooks Range — a seven-hundred-mile east-to-west
mountain chain spanning across northern Alaska and Canada —
is almost completely uninhabited except for the grizzlies, black
bears, caribou, moose, lynx, wolves, and wolverines that call
the mountains home. Swarms of summer mosquitoes, -60°F
winters, and government ownership of the land have slowed
the spread of human settlement and enabled the mountains to
maintain a primal "I'll kill you if you get lost in me" character.
And while the Brooks don't reach the towering heights of the
Alaska Range to the south, their desolate nature makes them
the sort of mountains that ought to be introduced with the
sounding of a Chinese gong and honored with reverent silence.

Paul and I, silent, felt unsettled and invigorated. It was as if
we had landed on another planet.

About an hour into our walk, we saw a trail of inch-deep bear
tracks next to the creek we were walking along. Paul and I
looked into each other's wide white eyes. We didn't speak out
loud, but it was clear that we were thinking the same thing: *We
need to get the fuck out of here right now.*

It was almost hard for me to believe the grizzly bear still
existed. It was an animal that would seem more at home in the

Pleistocene alongside our other extinct mammalian cousins like the woolly mammoth and saber-toothed tiger. Nevertheless, we were terrified about the prospect of running into one, as we were defenseless except for the ten-inch-long decorative knives that dangled on our belts (which we'd bought at a pawnshop in Fairbanks) and a canister of bear spray, which is high-powered Mace that has shown some effectiveness in repelling charging bears. Fears aside, some part of me harbored hopes of standing face-to-face with one. This is what I came for, after all: an adrenaline overload, a blow, a shock to my system — something that would charge every fiber of my body with screaming life; something that would scare the suburbs right out of me; something that would wake me out of my slumber and make me bellow, once and for all, "Holy shit. This is real!"

From the vantage point of a plane, one might wonder why the verdure of the tundra valley — this fertile plot of grassland — isn't dotted with thriving agrarian villages and cozy hamlets. From above, the land appears so flat and well trimmed — the sort of well-groomed lawn fit for picnics and pickup football games. Yet it's not until you're on the ground that you realize the arctic is carpeted with some of the most unforgiving terrain imaginable.

After spotting the bear tracks, Paul and I escaped the shadowy, tree-canopied creek and made our way toward the grassy hills so we could see all around us. En route, we trudged through mud-bottomed swamps; through dense, junglelike thickets of dwarf willow and alder; across spongy sphagnum moss; and over fields of tussocks, which were easily the worst of our hiking obstacles. Tussocks (or "nature's herpes," as I'd come to call them) are round, furry clumps of sedge that look like hairy green basketballs and are almost impossible to walk through without twisting an ankle and cursing maniacally.

Paul was so slowed by the tussocks that he couldn't keep up with me (not that I was managing them much better). Eventually, we lost sight of each other. *"Paul! Paul!"* I shouted. I didn't hear him shout back, so I retraced my steps and found him sitting on the tundra, rubbing his feet and moaning about blisters.

The terrain was unyielding, unmerciful, impossible. My one recurring thought: *People actually do this for fun?* Seeing Paul hurt was actually somewhat relieving because his injury now gave us a good excuse to quit and go home.

This was more or less what I'd expected. I'd figured nature was all about blood, sweat, and tears. It was about the hard lessons its travelers learned, as they did in Jack London novels. While I'd admired the works of the great American nature lover John Muir and, later, Henry David Thoreau, I never really understood their glorification of nature. Thoreau saw the world in the veins of a maple leaf, and Muir, it seemed, could find God in a mouse turd. Nature, to them, was transcendence, beauty, divinity. To me, nature was more like a football field or hockey rink in which games are won and lost. Yet, at the same time, I fantasized about experiencing nature the way they did. Maybe I'd undergo some sort of holy, transcendent awakening out here. Maybe I'd climb Blue Cloud and come back from the hike with a deeper voice and a newfound connection to spiritual realms. Squirrels and ravens would sit on my shoulder. I'd be able to predict weather from a lone gust of wind. I'd become one with my chi.

"So you're sure you can't make it?" I asked Paul in a tone of sly entreaty.

"Yeah, I don't think my feet can take anymore," said Paul with feigned regret but genuine embarrassment. "I think I gotta head back . . . What are you gonna do?"

I looked at Blue Cloud. The hills around it were green and bulbous, as staid and solemn as a shrine of plump, cross-legged bronze Buddhas. Behind them rose Blue Cloud. It thrust itself over the hills, puncturing the cloudless blue sky, a warped coal-colored arrowhead with veins of snow bleeding down its rocky grooves. It was miles away.

Unfamiliar with the sensations of determination that were colonizing my chest, the following words stumbled from my mouth awkwardly, in a half-statement, half-question sort of way, perhaps because I'd never had the chance to say anything cinematic before. "I think I'm gonna keep going? I'm going to climb this mountain."

Alone in the wild. I had no idea that the departure of a person could make me feel so different. With Paul gone, it was as if someone had suddenly cut the cords of the safety net that, to this point, had been hanging beneath me. I was now at the mercy of unfamiliar and abstract concepts: Nature, Fate, Destiny — it was they, along with my wits and a pair of trembling legs, that would determine whether I'd make it back alive. *This will be my great Alaskan adventure,* I thought. Oh, how I wished to tilt my head back and howl! And that's just what I did, unleashing a throaty, voice-cracking, off-key barbarian roar to the wild, man-less world around me. In mid-roar, I accidentally swallowed a mosquito and coughed, then continued on — now alone — lurching over tussocks and plowing my way through thickets.

The tussocks were everywhere. Every step demanded my full attention because each tussock required that I lift my leg extra high. After nearly twelve hours, the bottoms of my feet throbbed with prickly pain. It felt as if the soles of my boots were lined with thumbtacks. I hankered for a rest, but I was forbidden from slackening my gait because mosquitoes — thousands of them — would cling to my skin in swarms if I paused for even a moment. I began to feel a strange tightness in my chest. I became so sleepy I had to will myself to stay awake. My fatigue probably had something to do with dehydration, for I hadn't had any water in the last six hours because I feared I'd contract giardia from the creek below, as a ranger back in Coldfoot had warned.

But then I remembered I had iodine tablets in my pack, one of which could purify thirty-two ounces of water. I tore through my pack until I remembered that Paul had them in his pack. He had the matches and the compass, too.

Oh no . . .

I'd forgotten to ask Paul to hand over his share of our survival supplies when he'd left. This was the excuse I was looking for. Now I really could turn back and no one at camp would think the less of me. This was the responsible, practical thing to do. It was the right thing to do. Yes, I really must turn back.

But I couldn't. I wasn't sure why I cared so much about climbing this mountain, but I felt like something important was at stake. Maybe it was because I'd always felt so average: I was an average student, an average athlete, an average son. I couldn't have been more average. Climbing this mountain was my chance to start over. While this climb – like any mountain climb – was pointless (and probably of no difficulty to any seasoned outdoorsman), getting to the top of Blue Cloud suddenly took on an importance of mythic proportions to me. I pledged to myself then and there that, unless I thought death was imminent, I would not turn back.

I'd been hiking for nearly sixteen hours straight over a cruel and ruthless terrain, hauling a pack that felt as burdensome yet indispensible as a fallen comrade in battle. Each time I stopped to rest, I wasn't sure if I'd be able to get up, having perhaps depleted the last of my energy reserves. My thighs were wobbly sacks of water and the muscles of my calves felt so tight I worried my skin might split open and expose raw flesh like a pair of unforked hot dogs overheating in the microwave. My feet felt pulpy: I flinched before each step, awaiting the inevitable sting. My voice was hoarse from screaming to make unseen bears aware of my presence, and my shoulders were covered in blood because the mosquitoes would drink from my deltoids where my backpack stretched my sweatshirt tight against my skin. With each slap, I'd kill twenty at a time, smearing their jellied remains into the cotton.

Yet after each rest, I was able to get up and take a few more steps, and a few more after that. At some point, I'd wandered into that strange territory between my perceived limits and my actual limits – that stretch of land called the "unknown," a territory as wild and unfamiliar as the Alaskan country before me.

Climbing Blue Cloud was like climbing a hill of pennies. For every two steps up, I slid down one because the loose rocks would move beneath me, jingling down the mountain with each foot placement.

As I neared the top, the rocks got bigger and boulder-like; I could now use my hands to pull my body up. Yet each time I thought I was about to summit, I'd get to the top only to see that I still had more mountain to climb.

Finally, I arrived. I staggered over to a boulder, set down my pack, and spun around. There were mountain peaks and mountain peaks, as far as the eye could see. They were colored with puffy white lichen, neon-green moss, and the tinted leopard prints of cloud shadows. The landscape was a field of soft-edged flames, an armada of shark fins, a geological congregation of serrated ridges, grassy mounds, glacial moraines, and ice-cored pingos. It was so still and quiet and motionless, so unprofitable, so oh-so wild. How could such a landscape be so barren yet so appetizing to the eyes? I would have thought an endless vista of five-thousand-foot rock piles, bearing only the most resilient patches of moss and lichen, would produce sensations of disgust and revulsion. But oh no. The pasture may entice us with its fertility, and the city, its sophistication, but the Brooks' allure is in its desolation – it's a beauty that frightens and awes; it sets your imagination astir.

The cool mountaintop gusts ruffled my sweaty hair and momentarily dispersed the mosquitoes that pursued me, affording me a brief respite from discomfort. I thought about what I'd probably be doing at home in New York right now: maybe pushing carts or playing video games. Yet here I was now: a cartpusher in Alaska, a suburbanite in the wilderness, a stranger in this wild north country.

I didn't feel my chi. Nor did I feel like I'd won or conquered anything. Instead, I felt nothing but awe and deference and humility. I was *nothing* compared to all of this.

• • •

When Paul had left me, we'd agreed to rendezvous at our starting point: an abandoned mining camp called Nolan. When I'd told him to give me twelve hours to get up the mountain and back to Nolan, I'd seriously underestimated how far away the mountain was and how much time it would take me to do all that hiking. So when he drove my dad's SUV back to Nolan and didn't see me waiting for him, it dawned on him that it was possible that I might not ever make it back. He slept in the SUV, hoping that he'd wake up to me opening the passenger-side door, signaling my safe return. But I never did. He returned to Coldfoot and would head back to Nolan twice more. And each time that he didn't see me, he felt overcome with a guilt that brought him to his knees, where he'd pray for the first time in years. Soon, concerned coworkers began conjecturing what wild animal I was inside of. A local ranger caught wind of my hike, so he interviewed Paul to gauge my likelihood for survival and find out if I had any "suicidal tendencies."

"You mean to tell me you guys went out there with no map, no survival gear, hardly any food, and in jeans?" the ranger said to Paul, before shaking his head incredulously. "And what's this about not knowing how to set up a tent?"

The ranger clambered aboard his bush plane so he could search for me from above.

I began my descent. I knew I'd been hiking for a while, but I didn't know exactly how long because the sun hovers confusingly above the horizon at all hours of the day during arctic summers. The mountain was so steep that I had to keep my back against the mountain, carefully lowering each foot into piles of gushing scree. When a large rock fell out from beneath me, I slid twenty feet down the mountain, the rocks tearing a hole into the bottom of my pack and ripping open the seat of my jeans.

I yelled at a big brown rock that I thought was a bear, startled a group of nine snow-white Dall sheep ewes standing on a ledge, and walked beneath a lone gray storm cloud that followed me cartoonishly, soaking my sleeping bag and adding another ten pounds to my pack.

The pains in my feet began moving up into my calves, knees, and groin, then finally into my lower back. My legs were two heaving, sap-heavy tree trunks that I had little control over. I flung one in front of the other, hoping that my strides would resemble something close to walking. I didn't walk; I staggered. If there had been anyone out there to see me, they might have guessed that I was an intrepid member of the undead who'd set out to spread the infection to remote pockets of Alaska, or a courageous convalescent who set out to prove that he could, in fact, travel in the backcountry despite having previously suffered a crippling spinal injury.

I had been hiking, without pause, for what I think had been twenty hours. My mouth was parched, my shoulders sore from the backpack straps, and my feet felt like tenderized hamburger meat. I sat amid a buzzing galaxy of mosquitoes that plunged their proboscises through clothes and into flesh. I slapped body parts – arms, shoulders, neck, and buttocks – furiously and indiscriminately, performing a sort of demented dance, a madman's Macarena. I felt the eyes of the grizzly following me everywhere. But each time I looked behind me, it was never there.

I had never experienced such misery. Now I could walk only for a few minutes at a time before collapsing. I began to wonder – with no shortage of trepidation – not *when* but *if* I'd make it back.

After one such collapse, I slept for half an hour on a pile of rocks, turtling my head into my sweatshirt to hide from the mosquitoes. When I awoke, I scanned the terrain. Something seemed different. I reached into my torn back pocket to retrieve my map; said, "Oh no!" because it wasn't there; and sat down to scan the landscape once more.

I was maybe ten miles from where I began, except I had no idea where I was or where I was supposed to go. And because the sun never set, I couldn't tell east from west, or north from south. I had no food, no map, no matches, and no way to call Paul in Coldfoot or signal for help.

I thought of my mom and dad back home. They had no idea why I wanted to go to Alaska, yet they had been so supportive. How could I have been so irresponsible?! How could I bring so much pain to such sweet people?! I was alone in the wild, on the first hike of my life, and I was lost.

I had only one option: I looked toward the horizon, found a point that most looked like where I'd come from, and told myself that that was where I was going to go.

Because Coldfoot became enveloped in a blinding forest fire haze, the ranger decided to postpone his flight, and Paul started to talk to coworkers about organizing search parties.

But he didn't need to. After twenty-eight hours of almost nonstop hiking, I'd find a plastic wrapper, then a footprint, then the Caterpillar tracks, and then Paul in the SUV waiting for me.

When I was on the hike, I told myself that I'd never hike again. Yet as the days went by and the pains in my feet subsided, I began to look back on my little adventure with a hint of fondness. When it comes to memories, it seems we all have an editor within who will – if it'll make for a good story – revise the senseless into symbols, or rephrase miseries into warm memories. My editor would make Blue Cloud a chapter, a turning point, the happy pinnacle of my life – that time I discovered in myself a sense of determination that I didn't know I had in me. I went back to climb mountains in the Brooks again and again that summer.

Perhaps there's no better act of simplification than climbing a mountain. For an afternoon, a day, or a week, it's a way of reducing a complicated life into a simple goal. All you have to do is take one step at a time, place one foot in front of the other, and refuse to turn back until you've given everything you have. At the end of the summer, Paul and I drove back to New York as fast as possible so Paul could enroll in community college and so I could finish up my last year as a fifth-year "super senior" at

UB and get my degree. (I signed up for a fifth year because not all my freshman-year credits from Alfred transferred over and also because I wasn't in a hurry to leave school.) I came back confident and calm. My hair stopped falling out. The Tourette's and baggy eyes and voices had vanished. Yet, with every passing day, I began to feel the pressure of all the debt weighing me down.

Halfway through the fall semester, my mother called me down to the kitchen table. In front of her was a scattering of old bills and manila envelopes. When I walked in, she stared at me for a moment and said, "Ken, we need to talk," which usually meant that she needed to talk, and that the conversation would quickly become a theatrical dispute that would end with one of us storming out of the room and issuing the standard family rhetoric of accusing the other person of being insane, usually in the form of: "Are you insane?!" "You're insane!" or "Oh my god, you're insane!"

"We need to have a discussion about your loans," she said gravely. "Do you realize how much money you're going to have to pay back?"

"Mom, geez, don't worry. I'll be fine. I told you. I'll deal with it when I have to."

My mom being in a fevered state of worry was nothing new to me. Because of her training as a nurse, throughout my adolescence I'd been subjected to one outrageous medical evaluation after another. I tried not to clear my throat in front of her because I knew she'd worry that I had whooping cough. I'd always wear a shirt around the house for fear that she'd diagnose one of my back pimples as a case of impetigo again. If I showed her an unusual lump on my scalp from a spider bite, she'd silently appraise my deformity, her skin would turn gray, her eyes would water, and she'd think, *Oh my god, my son has three months to live.* After so many misdiagnoses, I couldn't help but become skeptical about all her misgivings. It became my instinct to counter her seriousness with playfulness, her pessimism with optimism, her insanity with inanity.

Unlike my mom, I'd hardly thought about the debt. I'd known that I had debt, of course, and that I'd one day have to pay it off, but that day always seemed so far away, as if it would take place in a second lifetime or a far-fetched futuristic world with flying cars, robots, and universal health care. I'd been putting off dealing with it from the moment I took out the loans. Debt, to me, was like death: I thought it was silly to worry about gloomy inevitabilities until I had to. My mother, though, couldn't get my loans out of her head. And each time she'd bring them up, I'd shrug off her warnings and shove worries to the back of my mind, where I'd keep them until the day I really had to start dealing with them.

"I don't know what you're going to do!" she said. "Ken, you're going to be $32,000 in debt. What are you going to do?"

"I told you what I'm going to do about my loans," I said. "I'm going to fake my death. You already knew that."

I was joking, but I really did—if just for a moment—consider the idea. I thought if I yanked out a few teeth, scattered them in my car, and burned it to a crisp before shoving it into Lake Ontario, then maybe the loan collectors would take me for dead. Without an identity, I'd have no choice but to melt into undocumented, under-the-table America—perhaps landscaping alongside Mexican itinerants or living on the second floor of a dingy pizzeria. Or maybe I could just skip the country and leave my debt behind. I'd go to some exotic archipelago, some lawless corner of Southeast Asia where I could embrace a life of crime. I'd start small, but over time I'd become kingpin of the region's drug trade. I'd develop a paunch, wear silk shirts, and maintain a tan. I'd be merciful, yet ruthless—a benevolent dictator of sorts, keeping obedient villages safe but crushing those late to pay their tributes with an iron fist.

Unfortunately, my mother had cosigned the loans with me, which meant the burden of paying them off would fall on her if I disappeared.

"Do you realize how much $32,000 is?" my mom asked. "You have two loans. The interest rates are 4.75 percent. If you don't make your payments, the interest will keep rising and rising, and so will your debt. Do you know what that means?"

"God, Mom, I'll be all right!" I said. "You're acting crazy again. I told you not to worry about it."

"I can't help but worry about it!" she cried. "What are you going to do?! Really, Ken . . . What are you going to do?" Suddenly, she put her head down on the table and wept. It was one of the few times I'd seen my mother cry. I looked on in disbelief.

What *was* I going to do?

"Really, Mom . . . Please don't worry. I'll pay them off."

Today, I realized, was the day I'd have to start dealing with my loans.

My debt, I decided, would be the next mountain I'd try to climb. It would be my Blue Cloud. It would be an adventure.

3

APPLICANT

August 2005-May 2006
University at Buffalo
DEBT: $32,000

I SPENT MY LAST YEAR of college trying to figure out
what I was going to do with my life. Throughout my education, despite becoming ever more aware of an unforgiving
job market and my impending financial crisis, I thought that
I'd be okay. More than okay, actually. I imagined that Fortune
would take pity and bequeath to me, and only me, a wonderful
job and modest salary. Maybe it would all go down at the Home
Depot, when I'd be loading drywall onto the bed of a customer's
truck, and the customer — who happened to be a philanthropic
billionaire (who, for whatever reason, did his home improve-
ment shopping in the sketchy part of Niagara Falls) — would see
in me some quality that no one else could, eagerly hooking me
up with connections in Washington to become a congressman's
trusty aide or inviting me to rehabilitate distressed seals in his
secret underwater dome.

But I didn't know anyone with connections: no philanthropic

billionaires, no seedy uncles, no former employers. Maybe I didn't know anyone, but I reassured myself that I was, in fact, an alluring job candidate. I could boast of a B.A. in history and English from a respectable college, a couple of internships under my belt, not to mention a long history of employment. *Who wouldn't want to hire me?* I'd failed to realize, though, that my credentials were identical to those of thousands of other job-seeking grads. Not only that, but my résumé indicated to prospective employers that I was capable of little more than low-skill, low-responsibility work that no one else wanted. Before I was a lodge cleaner in Coldfoot and a cart-pusher at Home Depot, I'd been a paperboy, a supermarket cashier, a public skating rink guard, a fast-food cook, and a landscaper. Between my liberal arts degree and my history of crappy jobs, I'd somehow made it through the first twenty-two years of my life without learning a useful skill.

I could, however, write reasonably well. I was a film reviewer for UB's student-run newspaper and, later, its Arts editor. Because of my experience with newspapers, I figured the print industry might be a good place for me to find my ideal job. So I applied to twenty-five paid internships (at $10 an hour) at newspapers across the country with dreams of becoming the next Bob Woodward, picturing myself as some investigative journalist working in a sweltering newsroom, bringing down corrupt politicians, and exposing the squalid working conditions of the city's immigrant community.

In due time, though, all twenty-five of my applications were rejected. And when the last rejection letter came in, I was only a couple of months away from graduating. I had no other job prospects except for mending the rip in my orange apron and heading back to the Home Depot, where, at best, I might someday be promoted to a department managerial position.

Somehow my brilliant friend Josh (my freshman roommate at Alfred) found himself in a similar situation. We'd kept in close contact over the years, sending e-mails to each other on an almost daily basis.

To: Ken Ilgunas
From: Josh Pruyn
Date: April 30, 2006
Subject: FUCKING JOB SITUATION
At the moment, my job search is the inescapable kismet of my existence. I heard back from both my interviews – both the predicted response. I probably sent out 5–7 apps in the past week, and have a total of 10 out right now. I wouldnt be surprised if I heard back from none of them. My ideal coaching/teaching job in Connecticut filled without even a word back from them. On my knees with my lips puckered, I sent another email asking to be reconsidered for the position. I've considered doing AmeriCorps, which after a term includes an educational award, but unfortunately I'd be unable to live on that sort of money due to my debt. Thats fuck ing pathetic. But I'll manage to pay my student debt one way or another.

Josh and I had known each other since we were six. We became best friends in the eighth grade for no better reason than our mutual interest in street hockey and video games. Our friendship began to develop at the age of seventeen, when we started e-mailing each other practically every day, using the recently discovered Internet to trade pictures of nude women. But over time, our e-mails became more intimate and substantive. Our subject matter expanded to politics, religion, worries, dreams, anything and everything. We didn't hold back. The more embarrassing, the more personal, the more self-admonishing – the stuff that a person feels most inclined to bottle up – was the very stuff we were most eager to share. Our e-mail correspondence was an interactive diary of sorts, a free therapy session, a cleanly scrubbed window through which we could view another human's soul.

Except for Josh's sometimes volatile temper and my tendency to revel in delusions of grandeur, we were incredibly similar, even more so now because the e-mails had a conforming effect. We were both liberals, shamelessly self-deprecating, and disdainful of school. In high school, we were losers, but because we were such boring losers, we were unworthy of the ridicule with which our betters honored our fellow, but more colorful, social outcasts. On weekend nights, we'd play computer games on the Internet together, commanding bit-sized armies till dawn while the rest of our classmates got high, drunk, and laid.

While we wanted girlfriends throughout our adolescence, we were both held back by several debilitating flaws. I was so awkward with girls you'd think I had spent my childhood locked in a damp basement; so unassertive that, on my high school football team, I let another player take my starting position as a defensive end merely because he wanted it; and so timid that our government and politics teacher once pulled me aside before class to ask me to speak, just so he could hear what my voice sounded like. Josh's drawbacks were more physical. He was unnaturally hirsute, shaving off a unibrow daily, coming to terms with random patches of back hair that escaped the reach of his razor, and contending with ass hair so long that its length – as he despairingly put it in one of his e-mails – could be curled "2–3 times around my finger."

Josh, though, had more than his fair share of redeeming qualities. He hated high school as much as I did but got high grades effortlessly. He had a gift for logical thinking that would make him, years later, the champion of the World Series of Euchre (a strategic card game popular in the Midwest and western New York). In college, Josh excelled, graduating with a 3.83 GPA and Phi Beta Kappa honors. He was the top runner-up for his senior class's Most Outstanding Student award, and his professors would cite his work in their own papers, one of them calling Josh the "most impressive student I've had the pleasure to teach." He was an idealist, fascinated as he was with the history of oppressed peoples in the Holocaust, Armenian Genocide, African American civil rights movement, and today's gay rights movement. He was inspired by Hannah Arendt's *Eichmann in Jerusalem,* as Arendt's words had shown him how corrupt governments are empowered by a complacent citizenry. He had dreams of joining the Peace Corps so he could help the poor and sick. It seemed he was destined for greatness.

Yet despite all the honors, accolades, and good intentions, Josh was in the same situation as I was. And even though Alfred gave him scholarships, his education still ended up costing him. He left Alfred with a B.A. in history and political science along with $55,000 in student loans.

After four years of school, he had no better idea about what he should do with his life than I did. So he did what many clueless young people do: He went back to graduate school to be part of a fully funded history Ph.D. program at the University of Delaware. But after one year – between second thoughts about grad school and worries about the interest that was stealthily accruing on his loans (which had leaped to $58,000) – he dropped out of school to find work.

Josh and I were nothing out of the ordinary. Like us, many students had spent their years in college thinking they'd get that well-paying, planet-saving job, even if they'd heard horror stories from recent underemployed grads. Those jobs, of course, no longer exist (if they ever did). By 2009, 17.4 million college graduates had jobs that didn't even require a degree. There are 365,000 cashiers and 318,000 waiters and waitresses in America who have bachelor's degrees, as do one-fifth of those working in the retail industry. More than 100,000 college graduates are janitors and 18,000 push carts. (There are 5,057 janitors in the United States who have doctorates and professional degrees!)

I'd heard of people who'd spent years, decades, their whole lives (!) paying off their debts, working eight hours a day, five days a week, fifty weeks a year, decades, lifetimes, epochs!

Before Alaska, I'd figured that I, too, was destined to be one of them. As I approached midlife, I'd begin to crave a red sports car or an affair to compensate for the youthful longings that I'd put on hold in my twenties. My life would be so monotonous and one-dimensional that I'd resort to aberrant role-playing sexual fantasies, delving into monthly feasts of feet, fetters, and fetishes behind my wife's back at "insurance conventions."

While hunched over reports, memos, and files, or in attendance at a grueling series of training sessions on diversity, sexual harassment, and office injury prevention, I'd remind myself that it will all be worth it when the mortgage is paid off and I can cash in and retire, finally taking to the life of the road, except now with an RV the size of an aircraft hangar and a prostate the size of a peach.

When I pictured my future self, I thought about my old coworkers and friends at the Home Depot: sad, tired, saggy-eyed souls who'd spend forty hours a week doing things they hated at a place they all wanted to burn down. Some were stuck because they had debts of their own, because they needed the health insurance, or because they needed the money to feed their kids. But it seemed they weren't all bound by these external constraints. Most were just too scared to leave. They tolerated the daily drudgery of work because dealing with daily drudgery was easier than quitting and doing something truly scary: sailing into unknown waters in pursuit of a dream.

I was different from them, I thought. Maybe I wasn't a year ago, but something in me had changed last summer in Alaska. I remembered my drive with Paul across Canada and how we felt like we could live our whole lives that way. The wandering life wasn't a ladder leading up an office hierarchy; it was a web of intersections that presented its wayfarers with U-turns, entrance ramps, and highway exits: chances to change the direction of their lives whenever they so pleased.

I thought I was made to live such a life – a free life: hopping trains, hitching rides, climbing mountains, traveling, wandering . . . I'd always felt uncomfortable when someone asked me what I wanted to be. But now I knew exactly what I wanted. I wanted the sensations I felt atop Blue Cloud and in the Brooks and on that drive to Alaska. I wanted to be a tramp.

I'd toil in Career World as long as I had to, but the minute my debt was paid off, I'd be gone. I'd tunnel my way under the businesses and institutions and corporations of the "real world" and come out, on the other side, with my body unshackled and soul unspoiled.

My goal was simple and straightforward: get the fuck out of debt as fast as humanly possible.

With graduation approaching and having no better idea about what to do for a job, I gave Coldfoot Camp a phone call. I'd heard a rumor that one of the tour guides wasn't coming back next summer, so I asked the camp manager if I could have the guide's job and let out a deep, thankful sigh when he said I could.

While the decision to go back to Alaska was made more out of desperation than desire, I justified that Coldfoot might be an ideal situation for paying off the debt. The nearest shopping center, after all, was 250 miles away (eliminating all temptations to buy stuff), there was no cell phone reception (making a phone plan unnecessary), workers got free room and board (thus eliminating food, rent, and utility costs), and my boss told me I now had the opportunity to make tips. If I stayed in Coldfoot for longer than a couple of months and flew up there rather than wasting lots of money on vehicle and transportation costs, then maybe – just maybe – I could take a chunk out of my debt, even though I'd only be making $9 an hour.

Despite the remote setting and low pay, I thought that getting rejected from newspaper internships and being forced to go back up to Coldfoot was serendipitous, a happy accident of sorts.

Tour guide. I liked the ring of it.

"I'm going to be a tour guide," I told my parents with a hint of pride.

"Oh, no you're not," they said, unimpressed. "You're going to get a real job."

But there were no real jobs, and it was too late anyway. Besides, I was excited about Coldfoot. There'd be no office cubicle for me. I'd be giving tourists van tours along the Dalton Highway and rafting trips down the Koyukuk River. I imagined myself leading intrepid guests through rigorous terrain, navigating through rapids, and dodging rock avalanches in the tour van on treacherous mountain passes. The remote setting, the adventurous nature of the job, and the small business that I'd work for – I hoped – would allow me to enjoy some semblance of freedom, despite being hamstrung by debt. Or that was the plan, at least.

I graduated on a May afternoon. When the director of ceremonies ended her speech, I had no desire to triumphantly fling my graduation cap into the air as a gesture of my newfound freedom. There was nothing liberating about leaving college: I'd ended one series of obligations only to enter into another. I looked with fear and uncertainty at the long road ahead.

The very next day, I hopped on a plane headed to Alaska.

4

·····················

TOUR GUIDE

Summer 2006–Coldfoot, Alaska

DEBT: $32,000

IF YOU WERE TO SMOOTH your hand over the arctic's contours, your palm would roll over smooth green hills, your nails would claw into moist fields of moss and sedge, and your fingers would run through bristly spruce groves. But you'd likely pause for thought when your fingertip landed on Coldfoot – a grotesque, relationship-ending mole, a protuberance that seems out of place on the arctic's otherwise unblemished body.

Today, Coldfoot is little more than a tire shop, a truckers' café, a small post office, a family-run bush plane air service, a ranger station, and, of course, the fifty-two-room motel that Paul and I cleaned the previous summer. Upon opening the doors to their rooms, the tourists – depending on their disposition – will either laugh good-naturedly or mutter obscenities about the austere accommodations: the sarcophagus-sized showers, the cigarette-burned carpets, the paper-thin wood-paneled walls, and the Nixon-era psychedelic-orange bedspreads.

The buildings, spread out and sprawled at strange angles, look like they've been haphazardly dropped from the sky. Surrounding them are the rusted relics of the town's industrial past: ramshackle mining equipment, abandoned trailers, derelict trucks. The motel and café are separated by a large, muddied, potholed parking lot where the semis – when parked parallel to one another – look like packages of frankfurters. The café, built in the 1980s by truckers who dropped off their empty packing crates, has a certain degree of charm, but doesn't exactly cancel out the camp's other industrial eyesores.

Indian and Eskimo tribes had roamed across the Brooks Range for thousands of years, but never were there permanent towns in the arctic until the Alaskan gold rush. In 1898, Coldfoot became one of many boomtowns where gold miners – too late to be among those who gathered the gleaming bounties of the Klondike – set out from to find mines in remote pockets of Alaska.

At its height, Coldfoot was home to seven saloons, a gambling hall, two roadhouses, a post office, and ten prostitutes. But by 1912 – after all the gold had been mined – the miners had disassembled the cabins and transported the logs 17 miles upriver to a new town called Wiseman, which is still standing today.

Coldfoot remained deserted until the mid-1970s, when the oil industry built a camp there to house workers who'd piece together the Alaskan pipeline, which stretches 800 miles from the oil fields in Prudhoe Bay on Alaska's northern coast to Valdez on Alaska's southern coast. To build the pipeline, the oil industry also built a 416-mile dirt road called the Dalton Highway that connected the city of Fairbanks to Prudhoe Bay. Coldfoot, situated halfway in between, was an ideal spot for a truck stop and is, today, one of the very few places where there are people year-round in the Alaskan arctic.

During the busy summer tourism season, Coldfoot's population triples to thirty-five. Typically, the crew is made up of college students and recent grads who work to pay tuition or their debts, but there are also a few middle-aged, carny-like drifters who make a life out of seasonal work, jumping from camp to camp every few months.

The day the summer crew arrived, we gathered by the river, built a huge fire from empty pallets and driftwood, and drank cheap whiskey straight from the bottle. Many of the workers from the previous summer had returned. There was Jordan, a thickset, dark-skinned thirty-year-old maintenance grunt who was born in India and raised in Alaska. He had twenty-two brothers and sisters (fifteen of whom were adopted) who'd all grown up in a tiny village to the south called Joy.

Ray, a dishwasher in his early thirties, was a Laotian American schizophrenic and alcoholic from Texas who lived in the room next to mine in the dormitory. At any hour of the day, I could hear him talking to himself through the walls in between swigs of Canadian Club. Every evening, in a drunken stupor, he'd kick open the hallway doors and sputter fake bullets that sprayed from a pair of invisible machine guns that he carried on each arm.

On the other side of my dorm wall was Avery, an eighteen-year-old waiter from the suburbs of Utah who spent the whole summer in the hazy daze of a marijuana high. Between him and Ray, Avery was the kinder, gentler mind-altered neighbor, whose guitar strums and bong bubbles canceled out the unnerving voices I'd hear coming from Ray's head.

Then there was Kerno the carpenter – a spitting image of Popeye's bearded, muscle-bound archnemesis, Bluto. Kerno brought with him his arsenal of automatic weapons, which he'd fire into a dirt mound behind camp.

Chad, a red-bearded thirty-three-year-old tour guide, was the much-loved, quick-witted Fonzie of camp and one of the few year-round Coldfoot residents. Behind the motel was the "dog lot," which was home to his thirty Alaskan huskies that he'd mush in the winter.

Natalia, one of the few girls in camp, was a twenty-one-year-old Ecuadorean student who worked as a lodge cleaner to make money for tuition. One day she knocked on my door in a white dress, asking if I wanted to practice my Spanish.

There were others in Coldfoot, but no one intrigued me more than an old man in his seventies. His name was James. He lived nearby and worked for the Bureau of Land Management, cleaning the outhouses on the highway. I'd spotted him walking across the camp in the early hours of the morning. He had a long, curly, white-as-death beard; a gaunt, wiry frame; and a spry, stilted gait. It amazed me that, for such a small town, I couldn't find anyone who'd spoken with him. I'd only heard strange rumors that he ate nothing but bizarre organic substitutes and lived in a copse of spruce trees on the edge of camp in his 1980 Chevy Suburban.

It was my job to drive the tourists up the Dalton Highway on a six-hour van tour through the Brooks Range. The tourists had signed up for an "Arctic Mountain Safari," but the name of the tour was more than misleading, considering that we didn't get out of the van except to use two outhouses on the drive north. I'd worried that the tourists would be disappointed with the extent of their "safari" – perhaps they'd imagined their tour led by a knowledgeable outdoorsman in a wide-brimmed explorer hat who'd ward off fanged mammals and identify different species of moss on daring hiking expeditions. But, much to my surprise, no one was ever upset with the tour. By the time the tourists got to Coldfoot, they'd been herded around so much on cruise ships, buses, planes, and trains that being herded around in a van was nothing new.

When they arrived on Holland America or Princess Cruises tours, they would stagger off the bus, seldom knowing where they were or who'd brought them. Once, a frosty-haired lady in her eighties pursued me threateningly with her cane, screeching, "Where am I?! *They never tell me where I am!*" Many were, as another guide put it, "forklift fat" – the sort of obesity that might require heavy machinery in funeral plans. On more than one occasion, I was asked to place my hands on a large Australian lady's bottom so she could be pushed into the van.

On the van tour, I'd tell them about the history of Coldfoot, the habits of the arctic's animals, and the geology of the mountains, which I'd read up on. We'd scan the hills and sometimes spot a caribou, but I'd always wonder: *Is this their idea of travel?* I thought real travel was about spontaneity and adventure. That sort of travel, though, was the last thing they wanted.

Yet I couldn't help but feel sympathy for them. Many only had a couple of weeks of vacation each year, and this tour was all they could afford or had time for. The retirees had worked hard their whole lives and no longer had the energy or strength to do more than this.

Two weeks after my first tour, I got a paycheck in the mail. I took a deep breath and opened it. It was for $300.

"Okay, not bad," I said to myself. "Just $31,700 more to go."

The season started slowly. In May, I was putting in a solid forty hours a week – mainly getting trained on how to guide a tour and oar a big blue raft. By June, I was off on my own, driving a fourteen-passenger white tour van up the Dalton, oaring a raft down the Koyukuk, and working a respectable fifty hours a week, which provided me with some much-desired overtime wages. By July, my workweek was up to sixty hours, and while I was happy to be bringing in the big bucks, I also began to mourn the loss of my free time, but I accepted it as a necessary sacrifice. By August, it was up to seventy. I wasn't just guiding anymore; several workers had deserted camp, so I was now making meals in the kitchen, hastily assembling extra coworker tent cabins for the next season, and helping the cleaners catch up on turning rooms. Some days work started as early as 5 A.M. and didn't end till 11 P.M. Alaska, which I'd once imagined as a refuge from the nine-to-five working world, turned out to be just as bad.

It feels almost blasphemous to admit hating work. It's true that people often complain about working twelve-hour days, balancing two jobs, or suffering through double shifts, but it seems our complaints are often just thinly veiled boasts about how busy our lives are, as if having no time for leisure, for a good night's sleep, or to do the things we actually want to do is some virtuous sacrifice we should all strive to make.

I may not have admitted it to anyone else – for fear of sounding entitled, or ungrateful, or whiny – but I hated work. I hated waking up early, hated taking orders, hated spending the great bulk of my time doing something for somebody else, hated how the hours would go by, hated how the days would melt into one another.

I didn't consider myself "above" work, but it just seemed so silly to have to work, perhaps for the next decade, and put all my earnings toward something as intangible, and clearly unprofitable, as my college education. And while I did think there was something crooked about the system – a system that charged unreasonable amounts of tuition to teenagers who only wished to better themselves and their society – I knew that I was responsible for taking out the loans and paying them back.

I despised having to work, now, more than ever, because my situation was so pathetic. After five years of college, two unpaid internships, and $32,000 of debt, I was just as unmarketable as I was as a teenager, doing the same sort of low-skill, low-responsibility, low-income work I'd been doing for years.

Although I was working for a wonderful company, with a wonderful group of people, my actual work didn't appear to have any of the positive qualities of good employment: It wasn't a way for me to climb the ladder to a better career; it wasn't helping me grow or develop; I wasn't spending my working hours basking in the warmth of job satisfaction by having spent my day making a useful product or providing a necessary service. I was in the tourism industry, so I was little more than a provider of luxuries and satisfier of desires. Ultimately, I was pointless.

I didn't see work – at least my line of work – as a virtuous undertaking. Rather, I saw it as nothing but a penance for my sins, for the profligate decisions I had made as a clueless eighteen-year-old. I thought of my job as nothing but frivolous toil – the only thing keeping me from living the free life I now dreamed of: of mountains and books and adventures and independence. I was ungrateful to have a job, yes, but I was grateful to have a life with which I might do better things.

I could have easily justified that I was lucky to even have a job unlike millions of other grads, or that I was better off than the twentysomethings who were starving, had AIDS, and were fighting a war somewhere in Africa, but I didn't want to make "the best of a bad situation." To make the best out of a bad situation seemed like an act of resignation. Instead, I embraced my bitterness and hatred and ungratefulness. Ungratefulness, I thought, was my only ticket out of debt and jobs like this.

And so, without time to visit the mountains or savor the books and lectures and higher things that kept me mostly sane in college, I relied on the blackout quantities of alcohol I'd lustily drink at weekly riverside extravaganzas – when we'd drink until we couldn't think of work or debts or troubles any more. In June, for my twenty-third birthday, one of the chefs, Karla, who'd given all the males in camp stripper nicknames (mine being "Kenny G String"), awarded me with a G-string she'd made out of velvet and muskrat fur that she'd acquired from a local trapper. After my eighth Miller Lite, egged on by a cheering crowd, I donned my present, wearing my G-string over my jeans like a washed-up arctic superhero. I can say from experience that when you wake up on a weekly basis with cottonmouth, a throbbing hangover, and a dead muskrat on your crotch, you can't help but question the direction your life is headed in.

I became obsessed with destroying what I thought was most constraining me. The debt wasn't a mere dollar amount; it was a villain that needed to be vanquished, a dragon that needed to be slain, a windmill that needed to be toppled. I thought of the debt as if it was the only thing keeping me from *really* living. It was the only thing on my mind. Nearly every dollar I made went toward my loans. I bought nothing and kept nothing in the bank. I squealed with pleasure when I tortured it with payments, like a sadist plucking legs from a captured mosquito.

On the drive back to Coldfoot on the van tours, we would stop at Wiseman, the historic mining village just thirteen miles north of Coldfoot. In the 1930s, Wiseman was home to 375 Eskimos and white settlers, but today there are only 15 people there, living largely subsistence lifestyles: hunting moose, caribou, grizzlies, and Dall sheep; growing vegetable gardens; foraging for blueberries and cranberries; and creating their own electricity with solar panels, as well as with wind and diesel generators.

Many of the cabins – made of stacked birch logs – are still standing from Wiseman's boomtown days. They are crowded by raspberry bushes or giant clusters of blue delphiniums. Rooftops are made of sod or shingled with old rusted tin oil containers or hidden under large solar panels. There are moose antlers hanging above doors; hunting caches, which looked like mini cabins on stilts; and vegetable gardens bursting with heads of lettuce and the greens of potatoes and carrots that would feed the townspeople for the long winter. My company paid one of the Wiseman locals to show the tourists his subsistence lifestyle.

Jack Reakoff was in his late forties, yet he didn't look a day older than thirty-five. He had a sturdy, frontiersman's carriage, and wore a necklace made of wolf teeth and a belt buckle made of Dall sheep horn. He'd lived in Wiseman ever since he was a boy and had grown up hunting, trapping, and fishing in the arctic's woods and rivers. As a young man, he attended the University of Alaska at Anchorage for a semester to study biology but quickly realized that there was no better classroom or teacher than that which he'd just deserted: the great outdoors. He came back to Wiseman and raised a family, embracing the arctic and the subsistence lifestyle that he'd never desire to desert again.

Jack would lead the tourists around town and invite them into his two-room cabin. Inside were mounted heads of Dall sheep and a grizzly. The walls and ceiling were covered with

pictures of his family and maps of Alaska. The room smelled of
soot, fried caribou, and body odor. Every day I'd listen to Jack
tell the guests about his lifestyle. I was his secret pupil, learning
everything there was to know about subsistence living in the
arctic. His was the most northern garden in America. In it he
grew hundreds of pounds of potatoes, cabbages, turnips, beets,
kale, and carrots, which he'd cover with clear plastic during the
cold summer nights. Underneath makeshift tarpaulin green-
houses, he grew zucchini, tomatoes, and peppers. He preserved
his winter vegetables in his refrigerator, which was a hole in
the floor of his cabin that remains 40°F year-round. He "killed"
trees that had reached full maturity, and then harvested the
wood on his snow machine years later when the logs were dry
enough for the stove. He ran traplines and sold furs. He hunted
and fished, and repaired his own machinery. To supplement his
subsistence lifestyle with money, he did these tours with Coldfoot guests for an hour or two every day during the summer.

I was having my first man-crush.

I envied Jack's lifestyle. It was the sort of lifestyle that makes
a man self-reliant, intelligent, strong; a lifestyle in which no one
ever had to think of the Dow Jones, the unemployment rate,
punch cards, or bosses. Jack's work and leisure, his toil and
enjoyment, were one and the same thing. His work was his life,
and his life was his work. Every day was a workday; every day
was a vacation. And he wasn't working in abstractions or pulling levers for some morally ambiguous corporation; his hands
were in the dirt, occupied as he was with the duties of feeding
his family and warming his home.

I could see that Coldfoot and Wiseman, as well as myself and
Jack, were polar opposites. One town relied on trucks to bring
food and fuel in from Fairbanks; the other created its own. One
town had wage slaves; the other, people as free as people can
conceivably be. Each town represented two types of living, two
types of working. They represented the life I currently had to
live and the life I wanted to live.

• • •

While I was getting paid plenty of overtime, $9 an hour was still my base wage, and I wondered if I'd made a mistake coming to Coldfoot. Perhaps if I'd looked harder, I could have found something that paid better and didn't demand so much of my time. I was, however, able to pay the debt off quicker than I originally thought, partly because my mother had offered to put my high-interest bank loan ($17,000) on her interest-free credit card. (My mother had perfect credit, allowing her such a card.) This meant that, for this loan, I'd no longer be paying back a bank — just my mom. This arrangement was advantageous to me because the interest on this loan would no longer accrue, and it wouldn't cost my mother a thing.

In addition to the $17,000 loan my mom had put on her credit card, I had another $15,000 government loan. Paying my two loans was a simple process. The payments for the government loan were automatically deducted from my bank account. Because I wanted to pay back my mom first, I paid the minimum on these government loans ($114 a month, about half of which would go toward accumulated interest).

Every week I'd mail my checks home to my mom, and she was supposed to put all the money toward my loans on her credit card. Our arrangement was working fine until one day, in the middle of summer, when I viewed my bank account on Coldfoot's shared computer and realized that she was putting the money not toward the debt but in my bank account.

I thought of my dollars like Spartan soldiers trained from birth to die in battle with debt, unconcerned with the likelihood of defeat. Yet here they were in my account, doing nothing, living it up like drunken sailors peeing on one another in port. I called my mother to find out what was going on.

"I see that you've gotten my checks," I said to her over the phone, trying to remain composed. "But they haven't been put toward the debt. How come?"

"Well, I put them in your bank account," she said sweetly, "because I thought it would be good for you to have some money just in case. You know, for emergencies."

"There won't be any emergencies. Can you please put it toward my debt?"

"No. You really need some spare money."

"It's my money, isn't it? Can't I use it the way I want to?"

"Ken, what if something happens to you? You don't have any health insurance. You know Dad and I don't have a lot, either."

"I don't care. Just put it toward the debt. I'm begging you. Please. Just put it toward the debt. You have no idea how much it bothers me. I know I sound crazy, but I want the debt gone more than anything."

"You're doing great with your payments, Ken. Why do you care so much?"

"I can't explain it. Just please put it toward the debt."

"No, you really need some reserve mon –"

"Put it toward the debt!" I screamed into the phone. "It's driving me *crazy!*"

5

....

GARBAGE PICKER

Fall 2006–Yukon River Camp, Alaska

DEBT: $24,000

ALONG WITH LEADING van tours and introducing the tourists to Jack in Wiseman, I also oared a large blue ten-person raft down the Koyukuk River for two hours at a time. The river was slow and gentle and easy to navigate, but it would pick up speed at the last bend in the river near Coldfoot.

On a trip with six retirees, I rounded that last bend where the river, predictably, picked up speed. I grabbed the rope, rolled over the side of the boat, and leaped out onto the gravel. The current was strong, so the boat yanked my legs into the icy water. But it wasn't a big deal: I was able to stop the boat from floating away and I didn't mind getting a little wet. Yet this caused something unexpected to happen. Just before the tourists went back into the inn, each of them – glancing apologetically at my wet pants – handed me a twenty-dollar bill. I stared at the wad of money in my palm in a state of awe.

I'd just learned an important lesson.

At the end of each rafting trip, I'd pretend that docking the raft was a hazardous, life-threatening undertaking. When docking it, I'd always make sure I got wet, sometimes even submerging my whole body underwater. "Everybody hold on!" I'd yell, as we approached the bank. "This might get a bit tricky."

From that point on, the tips poured into my pockets like the mighty Koyukuk. Each night, I'd bring home stacks of five-, ten-, and twenty-dollar bills, sometimes with a fifty shuffled in. I became an expert at spotting moose hiding in clusters of spruce trees. I'd laugh at the tourists' jokes, tell them about my Blue Cloud adventure, and hold my tongue when they were rude. Despite my timid nature, it turned out that I was an okay tour guide. And I could see that the elderly guests thought of me as a well-meaning and cash-strapped grandson. While I felt bad for taking their money, I knew I was giving them what they wanted at the same time.

Sometimes I'd come back to my room from a day's work with more than $200 in tips. I'd count the bills over and over, making sure all the heads were upright, before stretching a rubber band around each stack and hiding them under my mattress. By summer's end, I was sleeping on top of $3,000. Tip money not included, I'd paid off $8,000 of my debt.

While I'd made progress on my debt over the course of the summer, things weren't going so well for Josh. He was $58,000 in debt, most of his loans were given to him by banks that charged high interest rates, and, worst of all, he couldn't find work.

We'd heard of students putting their loans on deferment for a year or two to either go back to school or find a job, or because they were sick or just lazy. Whatever the reason, each story ended the same. Their debts — when they decided to begin paying them off — doubled. So if Josh let the interest accrue like other graduates had, his debt could get so big and unmanageable that paying it off might be impossible.

Josh, though, was hopeful at first. After all, he did wonderfully in college. He figured that with his B.A. and a year of grad school, he might be able to, at worst, land a respectable office job that paid in the low thirties.

At first he applied to jobs he thought he'd like, such as working for a medium-pay nonprofit or as a counselor at a school for troubled youths. After the first round of rejected job applications, he was rudely awakened to the fact that — even with his impressive academic record — he was a dime a dozen, just one of thousands of liberal arts majors who had to turn to corporate America to pay for their idealistic degrees. All prospective employers told him that he was either too unqualified or inexperienced. He had no choice but to lower his expectations.

> **To:** Ken Ilgunas
> **From:** Josh Pruyn
> **Date:** August 21, 2006
> **Subject:** Re: Josh's job situation
> Determined to find a job, I started looking through internet job ads like a frantic mother whose child wandered away in a department store. I dont even know what sites I was on, but the end result was harrowing: I applied to be a sales representative for a publishing company and a health care recruiting firm, a fraud analyst for Bank of America, and a financial advisor for Ameriprise. As I sit here now, I could barely describe what each position even entails. I hate banks more than anything. And yet I consider working for one? Combine that with the miserable, bottom of the food chain role in the organization I'd play, and I'm just disgusted at the thought of working for these companies.

He decided to keep looking for a job in accord with his ideals, so he moved into his parents' home where he could save on rent and apply for more work. Weeks later, he still had no job.

> **To:** Ken Ilgunas
> **From:** Josh Pruyn
> **Date:** September 15, 2006
> **Subject:** Coldfoot and future frustration
> I desperetly want to move on. I can't take it here, and my debt

feels like a raging cougar chasing me down while I just stand there tying my running shoes. I need to figure out the next stage of my life and I need to now. I dont think either of us has ever been in the situation where we had absolutely nothing lined up for our future — and its a terrible feeling. Add this into my debt and I'm anxious . . . And this feeling is almost constant.

And a week later . . .

To: Ken Ilgunas
From: Josh Pruyn
Date: September 23, 2006
Subject: Re: Coldfoot and future frustration
I need to fucking find a job and figure ouT WHAT THE FUCK I AM DOING WITH MY LIFE BUT I CANT FUCKING DO THAT FOR THIS REASON AND THAT REASON AND I'M READY TO GODDAMN EXPLODE FUCKKKKKKKKKKKKKKKKKKKKKKK KKKKKKKKKKKKKKKKKKKKKKKKKKKKKKKKKK

And a few days after that . . .

To: Ken Ilgunas
From: Josh Pruyn
Date: September 25, 2006
Subject: career frustrations
My friend, I am at a crossroads and I dont know where to go. I've been delaying and delaying applying to those jobs. Marketing Trainee. ugh. When I thnk that I've even been comparing it to the military . . . And in all honesty, I'm starting to think I'd rather join the Army than be a sales representative. I dropped the military idea to you before and I told you there may be a time when you needed to slap me for this . . . well now is the time because I can see myself researching that route in the next couple of days. Please get back with your thoughts on all of this ASAP . . .

Army? Josh in the army? Josh's hero was *Ralph Nader*. Josh going into the military was like a vegan working in a slaughterhouse, or a feminist on the pole. It wasn't like the country was in one of those rare periods of American neutrality. Hell, we weren't in just one war but *two*. Had it really come to this already?

What could I say? While I'd found a job and had been paying off my debt, I was in no position to give advice. Yet I was worried that he really would get stuck in some cubicle or, worse, shot down in a helicopter over Kabul.

I wrote Josh an e-mail asking him to come to Coldfoot. He read it and took a steaming hot shower. When he got out, he excitedly wrote me a response saying something along the lines of "Okay, you're right. What do I need to do?"

Unlike Josh, I had job security. Even though the tourism season was over, the truckers' café was open year-round, so I decided to stay in Coldfoot to be their night cook for the next eight months.

My boss, realizing that his trio of guides had been overworked during the summer, invited the three of us to join him for a fall weekend in the town of Valdez, where that year's Alaska State tourism convention was being held. I didn't see myself in the tourism business for long, but I figured a vacation would do me some good. As expected, there was fine dining, workshops, networking, and motivational speakers. I decided to attend the speech of a fifty-four-year-old Canadian motivational speaker, Bob Abrames, once a CEO of a Canadian travel agency and now the "World's Foremost Voyageur," as his promotional poster advertised. I had no idea what I was in for.

His head was spit-shine bald, and from his chin dangled a straggly, gray, Mennonite-long beard. Bob had the sort of stare that would burn holes through cotton. He had my full attention before he even said a word.

"Does anybody know what this is?!" He held up a long red scarf and draped it around his shoulder. "This . . . is a voyageur sash!" He marched from one end of the stage to the other. "This is a sash worn by the voyageurs!"

Canadian voyageurs, he would tell us, were like the United

States' version of mountain men. They were a group of people who traveled across Canada in birchbark canoes, transporting goods to the western frontier and bringing back fur pelts to the cities in the east.

"2005!" he screamed. "There's an article in the newspaper. And this article says, 'We're looking for nine people to paddle a birchbark canoe from Montreal to Winnipeg. You will live outside! You will not have tents! No sunscreen! No toilet paper! No toothbrush! One pair of pants! One shirt! You will eat boiled peas and salt pork! For one hundred days!'

"What do you think I said?" Bob asked rhetorically, pausing to look over the audience. "My god! Yes, I'll go!"

Bob spent the summer of 2005 living as a voyageur, starring in a ten-episode documentary called *Destination Nor'Ouest* that was a hit in French-speaking Canada. He lived exactly as the article described: paddling ten hours a day in a birchbark canoe, abstaining from modern conveniences, and wearing the clothes and using the gear of the eighteenth-century voyageurs.

He spent the next hour telling us about this voyage. He spoke of rapids, adventure, sacrifice, "fortitude in distress," glory. Glory, goddammit! I had no idea what any of this had to do with tourism, but I was entranced. At the end, he announced that he was embarking on another voyage the next summer. This time, he was organizing and leading an expedition of his own, and he was seeking three voyageurs to paddle with him. (My thought: *My god! Yes, I'll go!*)

But then I remembered my debt and how there was no conceivable way I could pay it off by next summer.

At the end of the speech, members of the crowd – equally affected by his words – surrounded Bob to shake hands and introduce themselves. Normally I'd be too paralyzed by my timidity to do such a thing, yet I found myself pushing through the crowd toward him.

When I got to him, I grabbed his hand and said, "Bob, I'm your man."

• • •

I came back to Coldfoot charged up from the speech. But because I didn't have the freedom to embark on journeys or voyages, I refocused my energies on paying off the debt. Get out of debt, then have adventures, I thought.

In Valdez, my manager mentioned that he needed help closing down one of his other camps for the winter. The Yukon River Camp (YRC), 120 miles south of Coldfoot, needed to be scrubbed clean. No one else wanted the job, so I told him that my friend Josh was looking for work and that he was willing to do anything for a steady paycheck. He agreed, and Josh flew up to Alaska. When Josh arrived, we bear-hugged each other and drove down to the Yukon River Camp.

The YRC is the sort of place that would lose to a Soviet labor camp in a charm contest. It would give Holocaust survivors flashbacks of bleaker times. It's the sort of depressing industrial work camp where you'd half expect to find, upon walking into your dorm room, your melancholic roommate Fred hanging from the ceiling by the elastic of his underwear.

A stone's throw from the 1,980-mile-long Yukon River, the YRC has a small café, a pipeline trailer converted into a one-story inn, and an abandoned tire shop that looks like a fitting stage for a horror movie massacre. The normally stunning landscape of the subarctic that surrounds the YRC looked grim and foreboding that fall, as the spruce trees had just been reduced by a forest fire to a field of brittle and blackened pencil-thin porcupine quills. The ten seasonal employees who lived there stayed in a regular-sized home (which would be quite cozy if it weren't for the adjacent diesel generator that ceaselessly blared outside). We Coldfoot employees jokingly referred to the YRC as "prison," partly because we were always worried about being sent there, but mostly because of the place's metallic austerity and the steely gaze of the seasonal coworkers—one of whom, Josh and I worried, might suck the very souls out of our breasts with their empty, end-of-times, coal-black eyes.

The YRC's claim to fame came in 2004, when a grizzly broke into the café during the winter (when the camp was closed down) and demolished everything inside. The bear devoured a bag of dried green peas and made a nest out of the T-shirts in the gift shop. In a desperate ploy to pay for the restoration, my boss tried to sell the pile of shredded tees to tourists by advertising them as BEARLY WORN.

Behind the motel were three giant barrels, each six feet tall and four feet wide, where the camp burned all of its trash: cans, food, plastic, everything. Around the rim of each barrel were gaggles of squawking ravens, shuffling their talons, eager for someone to throw away an uneaten hamburger bun.

We were brought in specifically to take care of the camp's "trash problem." First, we had to pour oil over the garbage and ignite it in order to reduce its weight and size. Second, we were to scoop the charred remains into black industrial garbage bags that we'd transport to a dump in Fairbanks. As we worked, the ravens, who tore through our garbage bags when we weren't looking, belligerently laughed at us from afar. We braced ourselves to yell back obscenities, but all that came out were hoarse, bronchial coughs.

The smoke caused an apocalyptic haze that reeked of flaming tires smothered in melting pig fat – the sort of smell that we presumed was shortening our lifespans and giving our sperm second tails. The toxins gave me a throbbing headache, but I wasn't suffering nearly as much as Josh. As I scooped a jumble of blackened cans into a garbage bag, I heard him wheezing. I looked over and watched him bend over and vomit. He wiped his mouth with the back of his sleeve, looked up at me, and smiled – a prideful, pathetic, pitiful smile – the same smile I had on my face. I knew exactly what he was thinking: *At least we have jobs.*

After several days of cleaning out the barrels, we spent the next few weeks brushing toilets, washing bed linens, and deep cleaning the kitchen. We hand-scrubbed the floors, sanitized the dishes, and scooped globs of auburn-colored, molasses-thick grease from every corner and crevice of the grill.

Even with the diesel generator blaring outside, sleep came easily. We knew that, with our hard day's work and $9 an hour wage, we had cut some meager but meaningful chunk out of our debts. Yet at the same time, the nature of our work made us feel pointless. I spent my working hours dreaming: dreaming of voyages, of the Brooks, of all the lovers I'd yet to meet.

When I was in front of the YRC by the Dalton Highway picking up trash, I looked down the road to the south – back to Fairbanks, to Canada, to the States, to a continent unexplored. Sometimes on the Dalton I'd spot a guy on his motorcycle or bicycle headed down to Tierra del Fuego, off the bottom tip of mainland South America. Some of them would spend years on the road.

Every half an hour, a truck would roll by. I wanted nothing more than to drop my trash bag and stick out my thumb. *I could just go,* I thought. I could leave it all behind: Coldfoot, the YRC, the work, the debt.

I stood there for a moment on the road. I stretched out my hand and extended my thumb, if just to see what it felt like. I looked south, down that quiet gravel path. A truck heading that way came rumbling toward me. *Why not?* I said to myself. *Why don't I just do it?* But it wasn't just the debt that kept me on the side of the road. There was something else.

When the truck came within eyeshot, I dropped my hand, picked up my garbage bag, and headed back to work.

6

NIGHT COOK

Winter 2006-2007–Coldfoot, Alaska
DEBT: $21,000

AFTER JOSH AND I FINISHED cleaning up and closing down the Yukon River Camp, we relocated to Coldfoot, where Josh would clean rooms for another month and where I'd be the camp's night cook for the rest of the year.

On our first fall weekend in Coldfoot, I took him out on what would be the first hike of his life as well as his first walk over elevated terrain. We strapped on a pair of small backpacks and set out to climb Twelvemile Mountain, a relatively short peak a few miles south of Coldfoot.

Once we crossed the river and climbed over the bank, Josh sprinted past me, running up the mountain at full bore, thinking the jaunt would be as easy as a leisurely jog around our suburb. Within seconds he collapsed next to a spruce tree, gasping desperately for air.

As we neared the top, we both – in a moment of juvenile camaraderie – stripped off our shirts and jogged along the hard, stony ridgeline, holding our arms out like airplane wings to keep balance. Josh got to the peak first. He placed his foot on the topmost boulder and screamed out into the vast panorama of blue-gray peaks.

I'd never seen this side of him. With his reserved disposition and reading glasses, Josh had always looked like your typical grad student. Yet now that he'd unveiled his grotesque shoulder hairs and was screaming at the top of his lungs – a jarring, orgasmic man-scream that echoed against the surrounding hills – I could see that Josh, too, had a wild side buried underneath years of civilization.

In the weeks that followed, Josh started going off on trips of his own. He climbed Blue Cloud and Snowden and Kalhabuk. He walked up Marion Creek, and when he got to a waterfall, he stood beneath it, letting the icy water pound his shoulders. He raised his arms like he'd just been freed from Shawshank, like he'd just been freed of debt.

But when winter came, there were no more rooms to clean, so Josh headed back to his parents' home in Niagara Falls and I stuck around in Coldfoot to work in the kitchen.

The day Josh left, winter arrived.

In the arctic, the seasons do not melt peaceably into one another as they do in other climes, giving each other handshakes, fond farewells, and see-ya-next-years. In the arctic, winter stands like a barbarian horde on the edge of town one day and ravages it the next. Winter moves in without warning, grabs summer by the belt loops of its cutoffs, and throws it out the door. It hardly gives poor autumn a chance to flutter its golden leaves, now caked in snow. The days rapidly become shorter and darker, and winter freezes, smothers, and colonizes everything with a bone-chilling, frostbiting, "I can't feel my penis anymore" cold.

From late October through March, the temperature rarely rises above freezing, and most days it stays well below zero. Negative 50°F is considered a cold day, but sometimes it can get much lower. In 1971, at a pipeline pump station south of Coldfoot, the station's thermometer read −81°F – the lowest temperature ever recorded in Alaska.

The arctic cold does not merely chill; it rips through layers of clothing, slices through skin, and bites into your brain, injecting your head with an ice cream headache that won't go away. Your nose and ears and cheeks turn pink, then a deadly white. It's a cold that cannot be warmed away; it's a cold that stays in your blood, frostily rattling against your bones like cubes of ice.

But it's not the cold that gets to most people in the arctic; it's the darkness. The sun rarely shows its face over the horizon, only casting a light glow behind the mountains for a couple of hours a day. From December 2 through January 21, Coldfooters never see the sun.

Five nights a week, I was the camp's line cook, working the 6 P.M. to 2:30 A.M. shift. I would prepare relatively simple meals — burgers, fries, fish sticks — never anything more complicated than an omelet. I'd flip burgers over the hot stove in the kitchen, some days in my parka because the café's heating system couldn't keep pace with the cold that would creep through the walls.

At midnight, I'd close the kitchen down by cleaning the grill, mopping the floor, and restocking the fridge. I'd scrub the bottoms of burned soup pots with a wire brush, dip my hand into sink drains to pluck out handfuls of slippery vegetables, and cram heavy black industrial trash bags into the Dumpster outside. Gone were the days when I'd worked seventy hours a week and pulled $100 in tips a day. While I was no longer paying off my debt at that staggering summer pace, I was still able to put my whole $300 weekly paycheck toward my debt. By the end of December, I'd paid off $11,000. I had $21,000 more to go.

Many of my summer and fall coworkers had left camp to head back to school or seek work in pleasanter climes. I'd said goodbye to Jordan, Kerno, Ray, and the rest of the fun-loving, hard-drinking, half-insane crew.

Coldfoot had only twelve winter jobs, but the manager had a hard time finding a winter crew because the camp received only a handful of résumés from mostly desperate applicants. It's no secret why so few applied. In the winter, not only is Coldfoot quite literally one of the darkest and coldest places on earth, but to make matters worse, there's also an almost prisonlike male-to-female ratio. When obese Eskimo women stopped to fill up their gas tanks, the males in camp would eye their robust flanks as if we were a pack of starving wolves. (But it wasn't any better for Coldfoot's few female inhabitants. The common Alaskan adage "The odds are good, but the goods are odd" holds true in Coldfoot.)

The new crew arrived at once. They were easily distinguishable from the previous set of workers. These guys were all full-time, middle-aged workers who'd been grotesquely bent by the winds of work. They had beady eyes; sinister, snickering hillbilly laughs; and faces marked with divots, cratered by hard times. They were the sort of men who – deprived of legal female companionship – wore faded briefs that were tiger-striped with skid marks.

There was Hal, a compulsive liar who told me he'd spotted frogs and snakes on his first outdoor arctic excursion. He said that he used to be a corporate trainer for Applebee's, that he co-owned a trading card company, and that he'd once been shot when on patrol in the army as part of a special forces unit – and while none of these was a particularly outlandish lie, I smelled BS right away. Despite Hal's many lucrative entrepreneurial engagements, he decided to take work alongside me as a fellow line cook (though, on his Myspace page, he listed his profession as "arctic bush pilot").

Then there was Walt, a gangly, slack-jawed, quick-talking forty-year-old maintenance grunt. He was a pleasant enough guy, but when Lucy, his hard-drinking Native American girlfriend, showed up, he turned dark and morose. While trying to sleep in my bed – a couple of rooms down from Walt and Lucy's – I'd hear wild, painful shrieks from their room before being startled by a loud thud against a wall. Perhaps they were the type who lovingly engaged in coitus while suspended from hooks in the air, but I got the impression that he was beating her up. In the early hours of the morning, Lucy, bedraggled, would knock on my door, waking me up to inquire if I had any beer.

There was Jonathan, a head chef and a cutter who left long red scrapes on the insides of his forearms. Boyd and Benji, a pair of alcoholic carpenters, moved in for a couple of weeks to help out with refurbishing one of the pipeline dormitories. They'd get inebriated after work, beat each other up, and kick each other's doors in (which, conveniently, they were able to repair).

There were also a few fairly normal people: Tom and Abbey, both college grads paying off their debts; Becca, a cleaner, who was saving up for school; and Jessica, a waitress, who'd just come from teaching English in the Czech Republic. For the most part, though, we were outnumbered by the crazies.

A Mormon punk rocker named Casey was one of our new lodge cleaners. We had a going-away party for Kerno – the affable Bluto look-alike – and afterward, in the middle of the night, Casey, who was upset with Kerno for some reason, came running at him screaming with his skateboard held high, ready to crash it over Kerno's head. From my room, I heard Casey screeching in pain in the hallway. I got out of bed to see what was going on. Kerno, who'd casually thwarted Casey's attack, had Casey in a headlock on the ground and was raining one blow after another onto Casey's screaming face.

"Kerno, what the hell is going on?" I asked, when I walked out into the hallway.

"This motherfucker ran after me with a skateboard!" Kerno screamed.

The only other person in the hall was Avery the pothead, who was taking a drag on his cigarette and watching the scuffle stoically, as if he were viewing something as serene and inconsequential as a family of gray jays enjoying a birdbath.

The fight ended without event, but every few days I'd hear more shrieks, thuds, and villainous cackles coming from the other rooms.

Avery, the 18-year-old waiter from the suburbs of Utah, was one of the few people left in camp who I knew wouldn't stab me in my neck while I slept. I'd never known anyone so mellow, so kind, so high. He was never not high. Let me rephrase: Avery was *always* high. Yet Avery, in ways, was more straight-edged than he had ever been. Working as a waiter in Coldfoot was apparently a huge step up for him. He started doing drugs as early as middle school, and apparently he could never get a job better than scrubbing down the walls of private viewing booths at a porn shop. While it seemed like many of his travails were self-inflicted, I couldn't help but feel sympathy when I found out that his father had abandoned him as a child and that his mom had begun having relationships with crazy lesbians. It was downhill from there, and anyone could see that Avery was destined for burned-out hippiedom. But just as he suffered from addictions like nearly everyone else here, he was different. He wanted to use Coldfoot as a place where he could reinvent himself now that he'd quarantined himself from all the bad stuff back home. He started by cutting off his long, greasy black hair and going cold turkey on pot. But between the drug withdrawal and the cabin fever, it wouldn't be easy for Avery. Once, he came running into the café during my shift, claiming he had a "renal." He called 911 in Fairbanks and they put him on hold for thirty minutes.

When he came back to the kitchen, where I was cooking, to order a meal, I said, "Hey, man. What can I get ya?"

"A bullet hole. In the head, please," he said.

"Dude, are you okay?" I asked.

"Good. You got off tomorrow?"

"Yeah."

"Well, if I'm still alive we should do something."

Avery was merely craving attention, so I gave it to him. For three hours I helped him apply online to be an Alaska state trooper, which was his dream job. Of course, no police department would ever hire Avery; he was the sort of hippie cops would love to come across at a peace rally so they could club an easy target over the head before flinging him face-first into the back of a riot van. I helped him apply anyway, as I was encouraged that at least someone in town was trying to improve himself.

Everyone else seemed to be decomposing: growing paler, sleeping longer, drinking more, and rarely, if ever, going outside. At first, winter merely brought gloom. Then madness.

The drifters—led by Lenny, Hal, and Casey—would get together each night, blare loud music in the dormitory, and drink hard liquor. Rumors of meth usage started to circulate around camp. At first, their bad habits were only mildly annoying and none of my business, but as winter wore on, things got scary. They started shitting on top of other workers' cars, pouring water on the huskies, who slept outside, when it was as cold as −20°F, and taking off on drunken nighttime joyrides in the camp's tour vans. Coldfoot no longer felt like a safe place to live.

"I'm literally worried about getting gang-raped," said Abbey, one of the servers, who admitted this to me after I told her that I was concerned about getting shot by the suicide bullet of my depressed dorm neighbor because we slept against the same wall.

"The shit we have to worry about . . ." said Abbey's boyfriend, Tom, shaking his head morosely.

On one of the drifters' drunken joyrides to Wiseman, they forgot to turn onto the Dalton and ran the van off a cliff and onto the frozen Koyukuk River. Avery and I thought that this was the end of our troubles because we were sure they'd be fired on the spot, but the camp manager had no choice but to keep them on. Camp would have to be shut down if it lost a fourth of its workforce.

The truckers were hardly any better. They'd strut their beer bellies into the café with a cowboy swagger and settle down in their section of the café where Fox News was on the TV at all hours. A few years back, under different management, the truckers were said to have coaxed the waitresses to their trucks for more generous tips. The saying went: "The waitresses would service them in the restaurant, then in the cab."

The café was haunted by the likes of "Big Dan," the fattest, most intimidating man I'd ever seen, whose arms were barnacled with open wounds; Pablo the Adulterer, whose every order seemed to insinuate something more; and Wesley, one of the highway's twitchy drug dealers, who'd hurriedly and apologetically cram Avery's latest stash into waitresses' aprons.

They were all so fat, and obscene with the waitresses, and picky, whining about "weak coffee" and "runny eggs," complaining that we didn't sell chicken-fried steak like other truck stops, and moaning that the food was making them sick, as if their usual fare of biscuits and gravy layered with bacon was supposed to make them feel healthy and chipper. Too lazy to pull over on their long hauls, they'd reputedly urinate into empty pop bottles at the wheel and then hurl the golden grenades onto the unsullied arctic landscape – a flagrantly insolent yet uncharacteristically dexterous skill that both appalled and impressed me. As their cook, I'd purposely hasten the onset of heart disease by slathering an extra coating of mayonnaise on greasy cheeseburgers and overloading their dinner plates with hillocks of golden-brown Tater Tots.

I'd come from a college campus populated with smart, ambitious, and well-meaning students and professors to this: a lawless bazaar of meth heads, alcoholics, and assholes who had me questioning the ideal of universal suffrage and who were compelling me to embark on my first foray into misanthropy.

I found it soothing to wash dishes, a winter duty that typically fell to the waitresses. The waitresses, happy to let me relieve them of their duties, were eager to reward me with a cut of their tips. I now made about $10 a day, raising my tip collection, over time, to $4,000, which I decided I'd keep for myself and not put toward the debt.

Natalia, one of the lodge cleaners who worked at Coldfoot during the summer, e-mailed me to ask if I wanted to visit her at her family's cow farm in Ecuador. Between the winter doldrums and itchy feet, I bought a plane ticket on a whim, spending half of my tip money on a three-week vacation. It was a pleasant trip: I petted llamas, saw my first-ever jungle, and drove through the Andes, and in honor of my visit, her family named a newborn *vaca* after me, calling it "Keña." But I felt guilty for spending money on something other than my debt. The plane ticket was a luxury purchase. I felt as if I'd cheated somehow; that I'd cut some corner that I promised myself I wouldn't cut. Worse, the Ecuador trip made me realize that I couldn't have "real" adventures so long as I had monthly debt payments to deal with. Instead, I'd have to sandwich my travels into three-week vacation blocks sandwiched in between lifetimes of work, just like the tourists who came up to Coldfoot. I wanted nothing more than to stay in South America. The jungle, the city squalor, the adobe huts—how it all captured my imagination. I wanted to keep moving, hopping from country to country, never stopping, always moving.

When I flew back to Coldfoot it was December 22, the winter solstice—the darkest day of the year. I sat in my room, alone, with a stack of books that I had no desire to read, beset by a constant, nagging, raging restlessness. Lying in my bed, I felt my skin crawl and muscles tingle. My whole body felt like it had been powdered with itchy pink insulation. I wanted motion. I wanted to move, move, move. Yet I was hemmed in. Stuck in Coldfoot. Stuck in this room. Stuck with all these miserable people. I looked out my window, glazed in ice, and all I saw was winter black. The winter blackness would cause the Eskimos to experience something called *perlerorneq,* or "the weight of life." They'd run out of their dwellings naked into the snow or eat dog feces. While it hadn't come to that, I couldn't bring myself to do anything more than lie in bed and stare at the ceiling. From there, I listened to the winds howl against the aluminum siding for hours, interrupted once by a thud against one of the other dorm walls, a "Yee-haw!," and a flurry of hillbilly high-fives.

In ways, we Coldfooters were similar to the Coldfooters of a century before. Like the miners, some of us were seeking fresh starts and new beginnings. Some of us came to the arctic for riches, others for glory. Some wanted adventure; others, refuge from an industrialized world. Many hoped to leave their troubles behind so they could reinvent themselves. But more often than not, we, like the miners, brought with us what we'd meant to leave back home. The people around me – some with genuine hopes of personal reformation – brought with them the drugs and alcohol and apathy that had plagued them for years. The town of Coldfoot was named for all the miners who'd come up there with sincere dreams and noble aspirations, but turned back home when it got too tough. They got cold feet.

I told myself that I was different, that I was nothing like these addicts and alcoholics. But in a way, I wasn't much different. I felt just as weak. I was thousands of miles from home, but I felt as contained as ever. I lived nearly the same sheltered, scheduled life I'd lived back in New York. And without the return of the sun to announce a new day, each twenty-four-hour period melted into the next. My life was an amorphous blob of flipping burgers, mopping floors, and sleeping. I was in a dream that I couldn't wake up out of. I was lost in a forest whose canopy concealed the sun, wandering in circles. Somehow I'd traveled four thousand miles yet had managed to bring with me the repetitious and ordered and cubicled life that I'd wished to leave behind. While I'd gotten myself out of the suburb, I couldn't get the suburb out of me.

My life was so monotonous, so goalless, so pathless. What was my purpose? To service truckers who worked for the oil industry? To pay off my debt? To work for years and years at shitty jobs and be thankful that I, unlike other people, had a job, or at least one not quite as shitty as theirs? I wanted more than anything to have something to work toward and strive for.

Something important. I longed to stop muddling around and dedicate my life to some high and noble purpose that would give it clear meaning.

I now understood why my old Home Depot coworkers looked the way they did: bored, tired, zombie-eyed. My life, like theirs, was so uniform, so one-dimensional, so unadventurous. I spent forty hours of my week doing things that didn't teach me anything new, that provided no variety, that tested no creative faculties. As a burger flipper, I was a specialist, a cog, an insect, hardly the human being that Jack in Wiseman was.

My journal was mostly blank. I wrote only about how I had nothing to write about. I couldn't remember the last time I felt any emotion in its extreme. How can you feel anything when every day is the same? I felt nothing. I couldn't remember the last time I'd cried or laughed or got steaming mad. It almost got to the point where I'd forgotten what these felt like. Give me anger and give me tears, but never this blank nothingness, this gnawing neutrality.

As expected, kitchen work was kitchen work, but occasionally, when the rare group of winter tourists made it up to Coldfoot, I had the good fortune of getting to lead an "Aurora Tour." On the tour, I'd drive them in the van to Jack's house in Wiseman, where we'd view the aurora borealis – the northern lights.

On my first Aurora tour, which I led in early January, I put on every heavy article of clothing I owned: a parka and a logger's hat that a departing coworker had sold to me for $20, a set of thermal underwear my mother had mailed to me, as well as gloves, wool socks, and bunny boots, which are large white boots made for polar excursions that Josh had found abandoned when we'd worked at the Yukon River Camp.

Coldfoot is one of the best places on earth to view the aurora. The camp sits directly under the auroral oval, which is where solar winds collide with and excite atmospheric atoms and molecules that cause a display not to the north – where most people see them – but directly overhead. Tourists, many from

Japan, would come up to Coldfoot during the winter months to see them.

I drove the tourists twelve miles up the road to Wiseman, where there were no outdoor lights to obstruct the views. Because the aurora would sometimes show itself for only a couple of minutes, the tourists would wait in Jack's cabin, where they stayed warm and drank hot cocoa while I stood outside waiting for it to appear. When it finally appeared, I'd run to the cabin, open the door, and yell for everyone to get out as quick as they could.

In all my winter gear, I managed to stay fairly comfortable in temperatures as cold as −40°F. I'd lie in the snow, looking up at the sky, waiting for the aurora to unfurl.

Nowhere had I ever seen a sky so full of stars. From my suburb in Niagara Falls – for the first twenty-three years of my life – I could make out only a few faint twinklings in the smoggy, light-polluted skies. But here, the stars gushed across the clear, clean arctic sky, a Gulf Stream of light that illuminated the rounded snowcaps on the Wiseman cabin roofs, making them look like squat mushrooms.

I felt a strange twinge of anger looking at the stars. It was as if I'd just learned of an inheritance that had been stolen from me. If it wasn't for Alaska, I might have gone my whole life without knowing what a real sky was supposed to look like, which made me wonder: If I'd gone the first quarter of my life without seeing a real sky, what other sensations, what other glories, what other sights had the foul cloud of civilization hid from my view?

We can only miss what we once possessed. We can only feel wronged when we realize something has been stolen from us. We can't miss the million-strong flocks of passenger pigeons that once blackened our skies. We don't really miss the herds of bison that grazed in meadows where our suburbs stand. And few think of dark forests lit up with the bright green eyes of its mammalian lords. Soon, the glaciers will go with the clear skies and clean waters and all the feelings they once stirred. It's the greatest heist of mankind, our inheritance being stolen like this.

But how can we care or fight back when we don't even know what has been or is being taken from us?

A pale green band appeared. It inched across the sky, a luminescent caterpillar slowly nibbling its way to the eastern horizon. Then several bands of light materialized – all parallel to one another – making it look as if the firmament wore a celestial comb-over. Those pale bands began to pulse. One ball after another would move down the green bands like a family of rabbits being digested by a python. And suddenly the aurora bloomed into full color. The sky lit up with spumes of reds, pinks, purples, and blues that swooped, twisted, and curled into each other. There was no sense, no order, no logic to the aurora's movement. It moved wildly and swiftly, changing into a different shape from one moment to the next. It was a glowing, throbbing, sashaying curtain of color, a Rorschach test that looked like whatever you wanted it to look like: a heavyset grizzly, a woman's hips, a highway climbing hills. The aurora was a powwow of ancestral spirits – writhing apparitions, conjured from the depths of a village bonfire. It was a desert storm, a million individual particles of light whipping over dunes in patterns that no human mind could comprehend or computer-generate. The aurora is alien and unworldly, but it does not frighten or flabbergast; it is a tranquilizer that sprinkles down onto its onlookers an opiate from the heavens. It puts you at ease. After a few oohs and aahs, Jack, the tourists, and I all turned still and silent, our heads tilted upward to space.

I was bearing witness to an ancient ritual that I felt I'd seen in a previous lifetime. I was being reacquainted with the images processed by a million eyes before me, reveling in the privileges of the great human experience. Money, prestige, possessions, a home with two and a half bathrooms – these aren't the guiding lights of the universe that show us our path. How can we dedicate our lives to such things when we can see the impermanence of everything above and below us, in the flicker of a dying star or the decay of a rotting log? The statues, the paintings, the epic poems, the things we buy, the homes we strive to attain, the great cities and timeless monuments. In time, they'll all be gone. And the names of the great kings and queens who shook the world will be forgotten, carried away like crumpled leaves from autumn limbs. Stare – really stare – into the womb of creation, and it will be impossible to dedicate your life to mindless accumulation. When you see the aurora, the only logical choice you can make is to spend the rest of your life seeking the sublime.

It never failed: When I'd gaze at the stars and the aurora, I'd see my problems for what they were. I always told myself that I'd been under the control of other forces: parents, school, work. And I'd convinced myself that my debt was to blame for everything, and me, for nothing (as if I had nothing to do with contracting the debt in the first place). I hated my job even though I worked for a wonderful company. And I told myself that, because of the debt, I couldn't travel, couldn't go back to school, and now couldn't even leave my room.

Part of me liked being in debt. Part of me even wanted to stay in debt, to keep going on random and expensive three-week trips to places like Ecuador so I could spend my hard-earned dollars on halfhearted adventures, instead of staying focused on what should have remained my true goal: busting out of the great American debtors' prison, steadily chipping away at its walls with each paycheck.

Part of me liked being in that position of submission, tied up in leather, willfully cowering beneath a ruthless whip-wielding Sallie Mae. Life is simpler when we feel controlled. When we tell ourselves that we are controlled, we can shift the responsibility of freeing ourselves onto that which controls us. When we do that, we don't have to bear the responsibility of our unhappiness or shoulder the burden of self-ownership. We don't have to do anything. And nothing will ever change.

I'd gotten too comfortable with my predicament. I was doing what the tourists and my Home Depot coworkers had done: I was placing the blame on my obligations and not on myself. If I was going to become the free person I wanted to be, I'd have to do more than pay off this debt.

When I got back to camp that night, I didn't know what to do with all this ambition. I needed some lofty goal to commit to.

I turned on the camp computer and did two things. First, I found ten graduate schools (six history Ph.D. programs and four creative writing programs) that I'd apply to. I did this on a whim, completely disregarding my goal of getting out of debt. I knew I needed to get away from Coldfoot and surround myself again with people who could help me elevate myself. I could justify that my year in Coldfoot was a gap year between under-graduate and graduate school.

Grad school was the logical, well-worn path I could take. But I also found the website of Bob the voyageur – the motivational speaker I met in Valdez – and learned that he really was going on a two-month-long canoe voyage across Ontario that sum-mer. I remembered how I'd approached him after his speech, yet I'd forgotten about it soon after. I sent him an e-mail, saying, in so many words, "I'm still your man."

MAINTENANCE WORKER

Spring 2007–Coldfoot, Alaska

DEBT: $16,000

I SPENT THE REST OF THE winter focused on accomplishing one of my two new endeavors: getting selected for the voyage or getting into grad school. For the voyage, I started a rigorous exercise regimen. Every afternoon, during the few hours of twilight, I forced myself to leave the toasty confines of my room to endure the cold outside. For grad school, I stayed up late, filling out applications and memorizing a stack of 1,500 flash cards with a vocabulary word on one side and its definition on the other. I was preparing for the GRE – a test required for students who wish to enroll in most graduate programs.

My social life all but vanished when Avery was asked by camp management to leave because he'd decided to start inhaling marijuana as if it were oxygen again. Because most of my other Coldfoot coworkers did nothing but drink and fight and shit on each other's cars, I kept to myself and took to my books. I read Barry Lopez's *Arctic Dreams,* Daniel Defoe's *Robinson Crusoe,* and Russell Banks's *Cloudsplitter,* as well as many of the works of Jane Austen, John Steinbeck, and Jack London. And then I picked up *Walden*.

When I read *Walden*, I found myself nodding to each paragraph, jotting notes in the margins, underlining whole pages. Thoreau gave me the words to describe what I'd felt for so long.

When he was in his late twenties, Thoreau moved out of Concord, Massachusetts, and into the woods next to Walden Pond, where he and a couple of friends had built a small ten-by-fifteen-foot cabin in which he'd live by himself for a little over two years. In his cabin, he embraced the simple life. He made his own furniture, hoed beans, and chopped wood to heat his home. Years later, he wrote *Walden*, a book about his experiences in which he championed the virtues of simplicity and damned the dangling carrots of civilization: the materialism, riches, and prestige that too commonly led men astray.

Some other notable achievements: Thoreau was an active participant in the Underground Railroad, using his cabin as a station to shelter escaped slaves; he went to jail for a night after refusing to pay his poll tax — an act of "civil disobedience" he committed to protest how taxpayer money was being used to support the institution of slavery and the crimes of the ongoing Mexican-American War; and he was an all-around renaissance man — a pencil maker, surveyor, naturalist, poet, carpenter, mason, farmer, gardener, and schoolteacher. He played the flute, held wildly popular melon parties each fall, and enjoyed dancing, ice-skating, and going on walks for four hours a day through the woods and fields of his native Concord. Historians claim he never had a sexual relationship, dying a virgin at the age of forty-four.

I was having my second man-crush.

Thoreau — coming from a family of pencil makers — went to college at Harvard in the 1830s. His financial situation, one biographer notes, was "perennially precarious." Like many students, he had to take off a semester to work so he could pay tuition. After he'd graduated and moved into his Walden cabin, he realized just how much he'd squandered as a student. "The student who wishes for a shelter," he said in retrospect, "can obtain one for a lifetime at an expense not greater than the rent which he now pays annually." It made perfect sense. I started to wonder why students – like myself – had put up with expensive food and housing when we could have devised more affordable, though rustic, living situations of our own. And I wondered why it seemed to be a prerequisite of life to have to work fifty weeks a year when Thoreau fed himself working only six.

Walden was published in 1854, but his iconic observation that "The mass of men lead lives of quiet desperation" seemed as apt an insight for today as it was then. He described how his fellow citizens ("serfs of the soil") would toil away at desks or on huge farms, hating every minute of it, just so they could live in large homes and wear fashionable clothes in order to impress their neighbors, who were also unhappily employed.

Thoreau made me feel like I'd been a sane man wrongly assigned to live in a madhouse. He became my guide, whispering wisdom to me through the walls of my cell, confiding to me that he's "convinced, both by faith and experience, that to maintain one's self on this earth is not a hardship but a pastime, if we will live simply and wisely."

It all made perfect sense. For nearly a year now, I'd been living in a small room with a lamp, a TV, and a single bed in the Coldfoot dormitory. It was the first time in my life that I didn't have a collection of electronic gizmos at my disposal. Between the free room and board at Coldfoot, as well as the absence of movie theaters, shopping malls, and other places where I'd usually spent my money, I had no choice but to save. And I had to be frugal. When I needed a haircut, a coworker cut it. When I needed an extra pair of pants, I grabbed some out of the cardboard box in which previous coworkers had thrown out their old clothes before leaving town. By having had to "do without," I discovered that I was, in many ways, better off.

While I began to feel rejuvenated by the longer days and warmer weather, Josh was in worse shape than ever. After working with me at Coldfoot and the Yukon River Camp the previous fall, he flew back home to western New York and found a series of seasonal odd jobs like delivering phone books door to door and helping as a UPS package handler.

To: Ken Ilgunas
From: Josh Pruyn
Date: February 23, 2007
Subject: Re:
Quick rundown on my current job . . . I'm working in the ghetto (emphasize ghetto) of Niagara Falls delivering UPS packages with a black driver named Leon. Leon is a character . . . almost a stereotype of a black man with money. He likes pimpin his car and overly large women. He is "workin" several women at a time – mostly overweight white whales. He showed me pictures on his phone of several of them exposing their breasts to him . . . He also showed me a video on his phone which appeared to be one of these girls sucking his dick.

When the holiday season ended, Josh was unemployed once again. This was nothing new, but things took a turn for the worse after he got a letter from his lender, stating that they'd miscalculated his debt. They said he owed $8,000 more, boosting his staggering $58,000 debt to a now crippling $66,000.

To: Ken Ilgunas
From: Josh Pruyn
Date: March 15, 2007
Subject: Josh loses it
There's only one human being who could possibly understand what I just went through and that is you. Directly after writing you my last email I opened a letter sitting next to me. Its contents sent me into a state of emotions that is hard to analogize to anything . . . At this moment you should be wondering what this letter possibly could have been about. My friend, there is only one element of society that could make me scream, yell, swear at customer service representatives, and spend 40 minutes jogging throughout the local neighborhood, cursing loudly as people gawked at me

from cars and porches who surely pondered what would make a young man jog through a blizzard. Yes, my friend, the letter was about my debt.

Even though Josh had consolidated his loans, he'd learn from this letter that he owed two financial institutions money. When Josh read through the letter sent by one of these institutions, he crumpled the paper in his fist and unleashed a horrifying, Schwarzenegger-esque "*Noooooooo!*" It was the sort of "*Noooooooo!*" that'd be appropriate to unleash only if: 1) You've walked into your kitchen to find your wife and child murdered by some evil cabal whose members gratuitously smeared semen into your mop-headed son's baseball mitt before placing it over his lifeless face; or 2) you've just found out you owe $8,000 more on your already-humongous student debt.

Josh would have to pay an extra $100 a month on his loans for the next twelve years, upping his payments to $700 a month. He didn't know what to do, so he busted open his front door and began sprinting through our old neighborhood in a blinding, eye-stinging Buffalo blizzard.

"I passed two cars," Josh continued in the e-mail. "Both of them slowed down and were looking at me with their mouths open. I passed some kids playing in the snow with their money [*sic?*]. The kids cheerily observed someone out in the snow, but their mother seemed shocked. I ran through the Meadows and through the development across Ferchen, backtracking at times to sustain the catharsis as long as possible. Eventually I came back, covered head to toe in snow, breathing heavily and thinking 10x clearer. $8,000. That is between three or four times the amount of money I earned shoveling burnt garbage and scrubbing toilets in Alaska for two months. I've paid about $1,500 of my debt so far, which felt like a satisfying start, yet as of today I owe more than I thought I did before I paid that much off. Another $8,000 . . . thats a summer in Alaska. All of the thoughts of how much $8,000 is make the larger number of $66,000 seem all the more dreadful."

He was able to land a three-month gig building homes for a low-wage AmeriCorps program in Mississippi, but that was the last of his adventures. Josh knew he had to find a permanent job, so he left our neighborhood in New York for Denver, where a close friend of ours offered him free rent until he got a job and his feet on the ground.

That spring in Coldfoot, new, younger, excited workers were coming in, and the older drifters were headed out. Hal the liar was fired. Casey the punk rocker was long gone. And slack-jawed Lenny got arrested by the local trooper after he beat up his girlfriend, Lucy, leaving strangling marks around her neck and blood oozing from one of her ears.

I took the new crew on long van rides to introduce them to the Brooks and organized weekly football games in the snow in front of the inn, and a few of the more book-loving workers and I even had a tea party on a frozen pond.

I was falling deeply, madly, "it hurts me in the gut" in love with Coldfoot and the arctic. And I was still knocking off $300 with each weekly paycheck. After a full year in Coldfoot, I'd made $22,000, which, except for my $4,000 in tip money, all went toward my debt. I had just $16,000 more to go.

When I calculated these numbers, I was in shock. While I only made $22,000 in a year, I saved 82 percent of it and could have saved nearly 100 percent if I hadn't spent it on my trip to Ecuador and other tiny luxury costs. Most would agree that $22,000 isn't a lot of money for a year's work, yet when you have no living expenses, it's a substantial sum. The key was room and board. Room and board was everything.

I punched a few numbers to see just how much money I'd saved because of the free room and board. I determined that if I were your average consumer, living your conventional home-dwelling, car-driving, supermarket-shopping lifestyle, I would have had to pay, over the course of a year, $27,540 of my income on living expenses (food, insurance, car payment, rent, etc.). That means I'd have to make $49,540 for the whole year if I

wanted to be able to save $22,000. So working in Coldfoot for a
year was *sort of* like getting paid $49,540 in any normal place. I
didn't own a car, a phone, or a home, but I was saving the sort of
money that a person of moderate wealth could.

I knew I'd leave Coldfoot with having learned at least one
thing of inestimable value: When we eliminate the high cost of
living, we can amaze ourselves with how much we're capable of
saving.

Normal Living[1] (per month)		Coldfoot Living (per month)	
Apartment	$1,000	Apartment	0
Utilities	$50	Utilities	0
Vehicle costs/repairs	$450	Vehicle costs/repairs	0
Gas and motor oil	$150	Gas and motor oil	0
Entertainment (TV, Internet, video games, travels)	$75	Entertainment (TV, Internet, video games, travels)	$25 (Netflix subscription and used books purchased online)
Miscellaneous	$75	Miscellaneous	0
Food	$200	Food	0
Car Insurance	$125	Car Insurance	0
Cell phone	$70	Cell phone	0
Clothes/household appliances	$100	Clothes/household appliances	$5 (Pair of gloves, secondhand winter clothing purchases)
TOTAL	$2,295 ($27,540/year)	Total	$30

1   The following "normal living" averages are inexact, based on anecdotal evidence
and statistical averages where I could find them. The "Coldfoot living" numbers reflect
a "reasonable minimum" that resemble my own expenses.

Best of all, I was in the kitchen less and less. More tourists began visiting around the spring equinox (March 21), when the aurora in Wiseman is most vivid, so I was frequently leading Aurora Tours. I also began helping the maintenance crew with renovation projects. I installed pipes and ripped out old carpeting, and for a week I got to varnish doors at the motel. Much to my delight, I learned that I'd be working alongside James, the seventy-two-year-old recluse of Coldfoot whom I'd yet to speak to.

This was a slow time of year for James at his usual job for the Bureau of Land Management, so the truck stop enlisted his services to refurbish one of their old pipeline dormitories. Despite his age and gaunt features, I was impressed by how fast James moved and how dexterous he was with his hands. It took me twice as long to do the work. I simply couldn't keep up, though I found that I could slow him down by prodding him with questions.

James, originally from Tennessee, had worked as a state trooper in Indiana. Upon retiring, he moved up to the arctic. There, without a place to stay, he decided to live in his 1980 Chevy Suburban for the next six years.

"How do you stay warm?" I asked

"I put a stove in there," he said. "You haven't seen it? The pipe is poking out the roof above the passenger-side door."

Later, we went over to the Suburban and he gave me a tour. The Suburban was about the size of a large truck. It was well kept, but its banana-yellow body had matured into a venerable manila over the years. And he wasn't lying: there was a propane stove inside. The propane tank itself was on a small trailer behind the truck — kept outside the vehicle for reasons of safety — but when it got really cold the tank would sometimes freeze, so he'd learned to bring a smaller tank inside.

He had pulled off all the side panels and ceiling upholstery inside the Suburban so he could insert fiberglass insulation. On the windows, he had attached some heavy-duty plastic for fur-

ther insulation. He cooked his meals on the propane stove and
slept on top of plastic containers in the back.

James's Suburban left me speechless. It was a mockery of
conformity, an affront to conventional wisdom, a symbol of
his complete lack of regard for the rules and norms and standards of our age. To me, the Suburban was freedom, unalloyed,
unadulterated, unblemished. Here – I thought, looking inside
his home – a man could be lord, monarch, ruler of his own tiny,
upholstered dominion.

"Do you even need money?" I asked. "I thought you were
retired."

"No, I don't need the money. I send it all to my grandkids who
are going through college. I love work. I'm seventy-two and I'm
going to be working till I'm a hundred. That's my goal," he said
with wide eyes, letting out a boyish high-pitch giggle before
coughing out some breathy grandpa-like chortles. "It keeps you
healthy. I couldn't be happier when I'm working."

"You don't get lonely here in Coldfoot?"

"No, I don't need to be around a bunch of people."

"But I heard you'd lived in Wiseman for a year."

"I only did that for a year," to which he added somewhat
scornfully: "Too many people."

"But James, aren't there only like twelve people up there?"

"No, there was more like twenty then."

"Do you care what your family thinks of you? Do they think
that you're eccentric?"

"I am an eccentric!" he barked, though without any hostility.
"Me and that Ted Kaczynski – the Unabomber – we got a lot in
common. I can see why he'd want to move out in the woods like
that. In fact, he had a lot more room out there in that cabin than
I have in my Suburban. The only difference between him and
me was: He was nuts!"

Most people would have considered James nuts, too, but I
thought there was something sagely about his advice. He represented everything I loved about the people in rural Alaska.

Except for the fact that many of them harbored tender feelings for ruthless right-wing politics – probably ruthless enough to make the libertarian, spear-wielding sand people of Tatooine grimace – I revered rural Alaskans for their independence, self-reliance, health, and happiness. And I liked the idea of work as a virtuous undertaking but knew I'd never be able to feel that way at Coldfoot, where I was doing little more than servicing the oil and tourist industries. I was beginning to feel grateful for my experiences, but I knew I needed a big change.

Each afternoon, when it wasn't completely dark outside, I'd head into an abandoned room at the end of our dormitory where there was a bench press, pull-up bar, and a few barbells. The room was unheated, so the walls were coated in frost like the inside of a freezer. I'd warm myself by using an old mattress as a punching bag, then do three sets of pull-ups, arm curls, and bench presses, all while wearing my facemask and logger's cap.

I knew I had little chance of getting selected by Bob to be one of his voyageurs, as there were only three spots and probably hundreds of applicants, but I thought I had an okay shot because I could boast in my application that I had a summer's worth of Alaskan rafting experience, which sounded deceptively badass.

Bob's website said that we'd have to undergo a fitness test to show him we could do one hundred consecutive ab rolls, squat thrusts, and lunges, among other physical requirements I was getting in shape for. I started going on jogs down winter mining trails every afternoon. At first, I was intimidated by the cold, but I learned that as long as I wore an extra pair of underwear and a facemask, I could stay warm at almost any temperature. I started testing my boundaries, running in temperatures as low as –10°F, then –20°F, and finally –30°F. Every minute, I'd have to take off a glove and pinch my eyelids, which would freeze together with frost, but I'd rarely be troubled by the cold after a couple of minutes of jogging.

The mountains, blanketed with a thick, creamy topping of snow, no longer looked so uncomfortably naked as they had in the summer. The howling of wolves carried across greater distances, and the trees all looked stolid and grave, like a phalanx of impassive Spartans with their spears held upright.

As the weather began to warm, I was beset with the aching need to spend every waking hour outside. With my remaining tip money, I bought a lightweight one-person tent for $125, a backpacking stove for $85, a water filter for $50 over the Internet, and a −20°F-rated sleeping bag from a coworker for $40. On my weekends, I'd go on long hikes, sometimes for several days on mining trails that had been packed down under the tires of heavy vehicles, making the hike manageable.

Things were going well, but every few days I'd get a rejection letter from the graduate schools I'd applied to. I'd applied to some of the best schools in the country, largely because they were among the few that offered free tuition and generous stipends, but also because I had it in my mind that someday I might amount to something more than an indebted line cook. I'd impressed one of my professors – the late Dr. Richard Ellis – with my senior paper on two early U.S. Supreme Court cases, so with his advice, I aimed high.

Within a week, my applications to Columbia and NYU were rejected. Next came letters from Yale and Princeton. And then the Universities of Pittsburgh and Wisconsin-Madison. Then the University of North Carolina at Chapel Hill and Notre Dame. Then the Universities of Minnesota and Miami.

I looked at all ten rejection letters on my desk. Goodness . . . *All ten?* Between these and the twenty-five newspapers that rejected my internship applications, I'd gone 0-for-35 in trying to become a "man of letters." And while each rejection stung, the one good thing about being continuously rejected is that your pride, over time, begins to callous over into a hard cheese, allowing you to comfortably walk over setbacks, even those that once seemed as harmful as a bed of glowing coals.

Strangely, I also felt an odd sense of relief with each rejection. I began to reexamine my original motivations for applying to school. While it was true that I sincerely wanted to develop my mind and become a better, smarter person, part of my decision had to do with fulfilling social expectations: with going to a big-name school to impress my friends and family; with climbing the socioeconomic ladder in hopes of reaching a comfortable life in a house similar to, but slightly bigger than, the one I grew up in. When I forced myself to think about my decision, I had a hard time imagining myself being happy stuck in a program, in one place, studying just one thing, for years. But none of this mattered. Deep down, I knew that there was only one application that mattered. And one day on Coldfoot's answering machine I heard a nasally Canadian voice addressed to me. It was Bob the voyageur. I called him back immediately.

"So you still want to join us this summer?" Bob asked.

"What do you mean?"

"Y'know. Do you want to come along with us on the voyage this summer?"

"Are you telling me that I'm going to be a voyageur?"

"Sure."

"Really?"

"Yeah."

I screamed and bounced across the room.

"Yes, of course I'll come!" I yelled.

I could have stayed in Coldfoot, guided for another season, and continued to focus on paying the debt, but this was an opportunity I couldn't pass up. I wasn't sure if I could physically handle the voyage, but financially I knew that my debt was in good order. I'd already paid my mother back and I was ahead several months on my government loan payments. I had the summer to myself.

I looked up airfare and was disgusted to learn that a one-way ticket back home was $600. "That's two whole weeks of work," I moaned.

• • •

In May 2007, on my last day at Coldfoot, I gave most of the stuff I'd accumulated away – some shoes, winter clothes, and books – and I mailed a box of my more valuable clothes and books back to my parents' home in New York. I crammed my remaining stuff – my tent, sleeping bag, and camping gear – into the large backpack that I'd first used on my Blue Cloud climb. I said good-bye to everybody, hugged Josh (who'd moved back to Coldfoot to take my position as a guide), and wished him luck with the mountains and his mountain of debt.

I looked over the Brooks Range one last time. They were alive, gushing water from melting winter snows that swelled streams and made the gentle Koyukuk River roar. An inquisitive moose watched me from a cluster of green spruce trees. The air felt practically alive. With each inhalation, my cells were charged with cold, eye-popping life. This place, once so new and exotic, had just started to feel like a place where I belonged. And while I'd been itching to get out of town for these past many months, now that I was finally going, I wanted nothing more than to stay. I remembered when I told Jack – the hunter from Wiseman – that this place was growing on me. He told me, "Everyone gets their home." I didn't understand what he was saying at the time, but now I found myself promising to return.

It was a mild, blue-skies spring afternoon. I was nervous, scared, but brimming (almost shaking) with excitement. I walked onto the Dalton and looked down the gravel road that stretched across a continent that – now at the age of twenty-three – I'd hardly gotten to see.

I heard a truck rumbling. It turned a corner and lumbered toward me, kicking up clouds of dust in its wake.

I thought of my year here in Alaska: Wiseman, Jack, the aurora, the winter cold, Thoreau, James's Chevy Suburban. Alaska taught me that anything was possible; that there are other ways to live, to work, to shelter oneself; that the cold wasn't so cold; and that – even in an age of inky oceans and suburban sprawl – there was still wildness. I thought of Blue Cloud and how, on it, I'd dipped my toe into the unknown. Today, it was time to submerge my whole body.

Up until now, I was never anything but a worker and a student. When I looked up at a dark arctic night sky, I thought I could be something else. I didn't want a job, a salary, a home. I didn't want to be a bolt in the consumer-capitalist machine. Or a boring Ph.D. student. When I looked at the stars, I could see my path. I wanted to be a comet hurtling through the sky, governed by no one's laws or expectations but my own.

I took a deep breath and clenched my fists, trying to gain control of my trembling hands. Where this truck was going didn't matter. All that mattered was that I didn't know. I slowly lifted my arm as if to wave. I turned my wrist, flashed a grin, thought of those three whispered words, and extended my thumb.

Part II

·························

TRAMP,

or
My Attempt to Live a Free Life in
Spite of Debt

8

....

HITCHHIKER

May-June 2007—North America

DEBT: $16,000

I WAS SITTING ON A COUCH in a one-story home in Teslin, a small Native American village of 450 denizens in the Yukon Territory. The homeowner, Tony – a middle-aged resident with narcoleptic eyes – was slouched in his recliner thumbing the remote control until he got to a show called *Kenny vs. Spenny*.

This was the third night of my hitchhike.

On his cushioned throne, Tony, with his tousled hair, baggy sweats, and Kokanee beer securely balanced atop his gut, looked like a Tlingit chief who'd fallen on hard times. What could have been a great man had become this: a lazy, drunken couch potato of no more consequence than a pile of dirty laundry.

I sat on his sofa, tense and rigid, my arms stiffly angled toward my knees. Charlene, a three-hundred-pound Native American woman in her late twenties, sat directly beside me. When she spoke, her hip rubbed against mine like a ship nudging against another at dock. Charlene had picked me up four hours before in Whitehorse – a city to the north.

"Listen, Tony," she said. "We need to get you out of here. You've been in this house for too long."

"I'm under house arrest," he said, while his eyes followed a police car that prowled past his window for the fifth time in an hour. "You know they're watching me, eh. I can't leave. I ain't goin' back to jail."

"Tony . . . We'll go to Watson Lake. We can party. You know you want to. Think about it," Charlene said, adding wistfully, "The gang, the booze . . ."

In midsentence, Charlene had lifted her hand from her thigh and suavely placed it on mine, just centimeters away from my penis, which had fearfully slunk into the folds on the opposite side of my jeans. It seemed that all of Charlene's inhibitions had been drowned under the life-ending quantities of alcohol she'd guzzled – a quantity so preposterous I would have considered it a cruel and unusual punishment if she hadn't relished every drop of it. She had started with a six-pack of wine coolers, then doused her throat with a bottle of tequila as if her tonsils were on fire, and now she lustily quaffed can after can of Tony's Kokanee.

"This is what we're going to do," she said, before taking another hearty swig. (Oddly, the alcohol seemed to have had no adverse effect on her speech or acumen; on the contrary, she was getting sharper by the minute.) "Tony, the next time the cops go by, you're going to leave the house, sneak through the woods, and wait for us on the other side of town."

Tony listened carefully, suddenly awoken from his evening torpor.

"And then we're going to pick you up and take you to Watson Lake," she said. "It's a good plan, eh?"

"Charlene, I can't drive," Tony said matter-of-factly. "They took my license away. And you can't, neither. You've already had six beers since you've been here."

For an ex-con, Tony – I was surprised and pleased to note – was exhibiting more than a good deal of prudence.

As Charlene ignored Tony's objections and continued to

unroll the blueprints of her plans, I took a moment to wonder: *How did I get myself into this situation?* I was in the middle of the Yukon, in the middle of two raving and possibly psychotic alcoholics, in the middle of the night.

"Tony, I'm not going to drive," Charlene said.

"Well, then who is?" he asked.

Charlene and Tony looked over at me and into my wide, frantic eyes.

So began our mission to traffic Tony out of Teslin.

I had given myself one month to hitchhike the five thousand miles from Coldfoot to my parents' home in Niagara Falls before the eighteenth-century voyage began. I decided to put my thumb out on the Dalton that sunny May afternoon, both to have an adventure and to transport myself home as cheaply as I could. Coldfoot had made me into a lot of things: a lover of nature, a competent cook, an able guide, and, most prominently, a cheap son of a bitch. Between the free room and board, the haircuts from friends, and the clothes I took from donation bins — I guess you could say I'd learned how to be a professional freeloader. And this, I figured, was a skill that might come in handy when hitching rides.

Before I left Coldfoot, I commissioned Josh with the duty of calling my mother if I turned up missing, which probably would have gone something like this:

Josh: Mrs. Ilgunas . . . uh, I have some news.
Mom: (silence)
Josh: Ken . . . Well, Ken kinda went hitchhiking and . . .
Mom: *What?!?!*
Josh: . . . he's missing in the Yukon.
      (My mom causes a supernova, there's a bright light, an explosion, and the world — as we know it — ends.)

I thought if I brought enough stuff, I'd be okay, even if I did go missing or got lost. So I jammed a week's worth of food into a gym bag and squeezed my camping supplies into my backpack, including my newly bought one-person tent, my sleeping bag, three pairs of jeans, three collared shirts, a rain suit, toiletries, a box of crayons, *Robinson Crusoe,* a six-inch hunting knife, my passport, $200 in cash, a black baseball cap that read COLD-FOOT CAMP in white letters, and three maps: one for Alaska, another for Canada, and a third for the Lower 48. I thought that if I slept on the side of the road and cooked meals on my back-packing stove, then I could travel across the continent practically for free.

But I also wanted to put aside a couple of weeks of my life when anything could happen: when I could mingle with ex-felons, have knife fights with grizzlies, or fall in love on the open road. I wanted to scissor-poke cowardice and reservation in the eyes and finally immerse myself in the unknown. And the road certainly was unknown to me, at least by way of hitchhiking. In my twenty-three years, I'd never even seen a hitchhiker. Not one. I wondered: *Could I – in this day and age – hitchhike all the way home?*

Today, the hitchhiker is little more than a dust-collecting cultural relic lodged in the back of the national memory, sitting beside the pioneer, pilgrim, hobo, and cowboy: each a symbol of freer times, but no more real than a child's action figure. While the hitchhiker can still be spotted on entrance ramps and roadways in Europe, New Zealand, and other places around the globe, in America, traveling with your thumb out is for the most part unheard of. I presumed this had to do with a few things: 1) the law, which prohibits hitchhiking in many states; 2) fear, which – thanks to B-rated horror movies and fear-mongering news media organizations – makes us think that if we hitchhike we'll be raped, murdered, and mutilated (though not necessarily in that order); and 3) too many young people have jobs they are unwilling to leave, either because of debt or because they didn't want to lose their health insurance.

Luckily for me, I no longer had a job tying me down, nor

was I going to let reasons one and two stop me from having my adventure. But I was definitely scared of what was out there. Like many members of Generation Y, I'd suckled the milk of paranoia from the teat of fear from the get-go. I grew up not just worrying but *knowing* that I'd one day be molested. As a boy, whenever I exhibited the slightest hint of melancholy my mom would check in on me and ask, "Ken, has someone been . . . *touching you?*" I was told that if I was approached, offered candy, or simply looked at weirdly by a stranger, it was more than likely that he was hankering to "touch me." I grew up thinking, *Why do all these people want to touch me?* (Which is a worry that, unfortunately, vanished as soon as I wanted people to.)

Any Halloween candy that looked like it had been tampered with was thrown away for fear that someone had injected AIDS into it. My brother and I weren't allowed to ride our bikes to the mall. Or the convenience store. Or across any busy streets. We were contained in our suburb the same way the townspeople in the movie *The Village* were contained in their community. Just like them, we had culturally prescribed monsters lurking on the edges of our neighborhood, keeping us in, keeping us safe, keeping us scared, keeping us bored.

As a young adult, I'd heard all the same warnings over and over again. I was reminded constantly that "it ain't how it used to be," meaning that the roving bands of rapists, child molesters, and face-masked henchmen out there were a product of the twenty-first century, and that while it might have been okay to do something adventuresome yesterday, it would surely be insane and suicidal to do so today.

After my year in Coldfoot, I was still as paranoid as ever. Yet I had a sneaking suspicion that this was all wrong. It was time to find out for myself, I thought.

On my first day – after I stuck my thumb out on the Dalton – the semitruck coming toward me lumbered past, causing me to turn my face so I could protect my eyes from the gravel and dust swirling in the truck's wake. I stood there for another twelve hours and watched seventeen trucks go by. (*Maybe it's not possible . . .*) I headed back to Coldfoot to reorganize and rethink my strategy. The next morning, I persuaded one of the truckers at the café, who was heading southbound, to help me get started on my journey home.

Dirk often frequented Coldfoot on his long hauls up and down the Dalton. He was in his late thirties. He wore a dirty ball cap, an unbuttoned flannel shirt, and a pair of oil-spattered jeans. He had a convict's goatee and a confident demeanor that seemed in sync with his filthy ensemble.

As I kept in stride with his caffeinated gait from the café to his truck, he explained that he needed someone to talk to so he could make it all the way home to the town of North Pole (almost a 270-mile drive) without falling asleep.

On the road, Dirk talked a mile a minute, detailing everything from his family history to his criminal past. (He confessed that he was wanted for "minor" crimes in two states, as well as murder in a third.) He told me about everything: his trucking company, his family, his love for automatic weapons, as well as his distaste for Subarus and the theory of evolution. Everything. He described his wife's pubic hair in exquisite detail. He told me how much he loved his children and wife (except for issues related to the management of her nether regions), and about a period of his life when he was a self-described "man-whore," luring strippers into his truck where he'd warn them, "If yur gettin' in my truck, yur gonna fuck" – a message that he thankfully refrained from relaying upon granting me admittance.

He also, in more sullen tones, told me how – when he was a kid – his father had molested his sister, and how he'd just recently told his father that he'd kill him if he ever caught him alone with his children. I didn't know what to say, but he chirped, in a more cheerful timbre, "You know what I'm thinking about? Rat-at-at-at-at-at-at!" He sputtered this while miming a 1930s gangster shooting a gun. Allegedly Dirk had bought a tommy gun for $1,300 that had come in the mail while he was on the road. "When we get home, we can test it out at the gravel pit."

After seven hours and 270 miles, he brought me home with him, where he introduced me to his beautiful wife, his nine-year-old son, Kevin, and his six-year-old daughter, Kayla. I made friends with their Saint Bernard and played tag with the kids, and they invited me to stay for a cheese and crackers dinner.

Afterward, Dirk assembled the tommy gun, and we took a ride to the gravel pit. The thought of shooting a gun had all my old reservations resurfacing.

Dirk took a few shots and said, "You wanna give it a try?"

I pulled the trigger once. A delightful shockwave spread across my chest. *Okay, that felt pretty good.* Then I held down the trigger, firing what felt like a hundred bullets in a matter of seconds.

This was perhaps the greatest day of my life.

The next day I got a ride with Jim, a hunter who lives in the bush. "You can't never beat a female in gun shootin'," he said. "You give a girl a gun the first time and odds are she'll shoot better than most any guy." He dropped me off in front of a school in the native village.

As I sat with my cardboard sign to Tok, the next Alaskan town on the road, a couple of girls in their upper teens greeted me with a "Hi there" and coquettish giggles. Despite the pattern of middle-aged men I had had for drivers so far, I harbored fantasies of having a subarctic tryst with a nubile young driver in a tight tank top and jeans — one who'd love me one night and leave me devastated the next. No longer at the mercy of the Sacred Schedule, I meant to strip off order, inhibition, and restraint and fling them into the air, like clothes in a moment of passion.

Lenny, a heavyset, middle-aged native, wasn't the dark-haired beauty I dreamed of. As I settled into the passenger seat of his battered pickup, I detected something off-putting in his wide smile and slurred pronunciations. I buckled up, swallowed my fears, and hoped he was merely dazed by a mild hallucinogen.

He was heading to Tok to get his tires changed, as he was giving his truck to his son as a graduation present. A strange gift, I thought, because his son had just "beat the livin' shit" out of Lenny – as Lenny put it – when they were both under the influence of meth. Perhaps Lenny deserved it. He told me he hadn't been much of a father figure. In addition to admitting to me that he still dabbled with meth, he mentioned nonchalantly that he "just got out," which I took to mean "of prison." I watched him sip from a can of Sierra Mist, and I wondered what squares of the periodic table he might have sprinkled in.

As we headed down the Alcan Highway, lined on each side with dense spruce forest, Lenny recounted stories of his eighteen-month stint in a federal penitentiary for killing a bald eagle. When I questioned him about what seemed like an awfully rigid sentence for killing a bird that was no longer on the endangered species list, he clarified that his sentence was lengthened because of a long list of DWIs, thereby confirming the worst of my fears.

First an ex-convict. Then a hunter. Now I was riding with an ex-convict/hunter with a penchant for nose candy and swerving over yellow lines. We were going about 60 mph, and I estimated that if I opened the door, exited the truck, landed on my shoulder, tucked, and rolled, then I might have a better chance of surviving the fall than I did in the truck with Lenny.

I stayed in, though, and he wished me well after letting me loose in Tok.

That evening, I slept in my one-person ultralight tent on the side of the road. I woke up the next day, cooked a pot of oatmeal with Craisins, and made a giant cardboard sign to Canada with a red maple leaf in the center. To the south, the tide of mountain peaks were frozen white, though everywhere else – the forests, the rolling hills, the tundra plain – was colored with a vivid green, for the land's white winter covering had just been melted by the warm breath of spring.

• • •

After another ride, I made it across the Canadian border into the Yukon, where I found myself sitting on a curb outside White-horse, a quaint subarctic metropolis in the middle of nowhere. It was there that Charlene pulled over for me in her sedan.

She was headed to a baby shower in Teslin, one hundred miles to the south. Because she was hankering to start "pregaming," she handed me the keys so she could drink. I thought there was something more than odd about giving a complete stranger the keys to one's car, but I was happy to keep heading south and fine with her drinking so long as I was doing the driving. And it would have been a pleasant ride if not for the fact that Charlene insisted we keep the windows rolled down.

"Are you sure you're not getting cold?" I asked, shivering feverishly.

"I'm fine." She giggled. "I'm a Yukon girl."

She was in the middle of telling me about her job as a social worker when she exclaimed, "Wait, wait, wait. Turn here at this campground. I got some friends I want to meet."

This is it, I thought. This was her plan all along. I pulled into the campground resignedly, knowing that this was when she and her cronies would toss me out of the car, steal my gear, rip off my clothes, and jam some noncylindrical object into me while laughing and giving each other high-fives.

But after a trade of hellos, that was it.

We arrived in Teslin around midnight. Lanky men were strutting through dimly lit streets openly drinking cans of beer and screaming obscenities. Because I was in a foreign place around people I didn't know, I wanted to get away from Teslin and Charlene, though I felt obligated to chauffeur her around town.

Later, when we visited Tony, and when Charlene proposed that we traffic Tony out of Teslin, I thought I'd tested my luck enough for one day. When we went back to the car I grabbed my backpack and said, "This has been wonderful, Charlene, and I really appreciate the ride, but I really should go my own way at this point. I'm sorry I can't help you out with Tony." She begged me to stay, but I found myself backing away step by step, until I finally turned around and briskly walked down a dark road lined on each side with the bone-white skeletons of birch trees. I came to a river and set up my tent on its shore.

As I shivered into my sleeping bag, I thought about calling it quits. In just a couple of days, I'd been in cars and trucks with ex-convicts and killers, alcoholics and addicts. All my drivers had devastating stories to tell. They had shared tales of poverty and pain, rape and child abuse. Maybe this place really is full of bad guys. Maybe everything I'd heard was true. Maybe I was better off heading back to Whitehorse and taking a flight home.

But then I thought about how I'd yet to see evil. In fact, despite all the horrific stories, I'd seen nothing but kindness. I needed to continue on.

It didn't take long for me to get good at hitchhiking. With a box of Crayola crayons, I'd draw big family-friendly signs on giant rectangles of cardboard that I'd get from gas stations and restaurants. I'd pick the best places to stand: entrances ramps, rest stops, and highways where vehicles would have plenty of room to pull over. I made sure to stand in areas where the traffic moved slowly enough so that drivers could inspect me and see my sign. On my backpacking stove, I cooked oatmeal in the morning and a Mountain House meal at night (freeze-dried food that just needs boiling water added). I slept soundly in my tent in forests or alongside highways. I kept my appearance neat and trim, showering at truck stops and shaving in gas station restrooms. I wore jeans, a collared shirt, and a baseball cap, and I flashed a meek half-smile at all passing cars. Most times, I'd wait an hour before getting a lift, sometimes two. Other times it took a lot longer. But I never got stuck.

After my night with Tony and Charlene, I sat next to Dennis, an uncomplicated, plainspoken trucker who took me the nine hundred miles from Teslin to Prince George, British Columbia. George, a Catholic priest with a frosty, neatly trimmed beard heading to Ashcroft, BC, told me about his hitchhiking exploits when he was a lad my age. Four teens on their way to Vancouver spoke of pot and sex the whole ride. I was stuffed in the back of their crammed car between two blondes; the one with milky skin asked me if I'd pretend to be her boyfriend.

Keith, a retired machinist heading to the U.S.-Canada border, told me, "You're not worth a dime until you make your boss a dollar." Kevin, employed by the nuclear power industry, told me the world was going to end in 2012.

In Washington State, when I asked Bob and Esther, an eighty-year-old couple, why they picked me up, they told me they were "some of the few left who still like to trust people."

John, a shirtless nineteen-year-old, pulled up in a rundown 1980s sports car in the town of Omak, Washington. He was thin and muscular, just out of basic training and on his way to Iraq, where he'd serve as a scout. He joined the army because he'd blown $10,000 on pot and had to pay it back quick.

"I told my recruiting officer that I want to get in shape, fuck up and kill a lot of people, and come home."

"Is that exactly how you put it?" I asked.

"Yeah," he said matter-of-factly.

I made signs for Wenatchee, Rock Island, Kittitas, Yakima, Richland, Kennewick, Pendleton, and La Grande.

My SOUTH sign stopped working on an entrance ramp in a sleepy farm town called Kittitas in the state of Washington. A man who introduced himself as Juan Hernandez—a Mexican immigrant with a contracting business in Yakima—saw me and decided to pull over, even though he wasn't heading in my direction. He took me to a Wendy's and, despite my objections, bought me a hamburger and fries, which he watched me eat. He spoke in broken, hard-to-understand English, but his passion for his god and his America was palpable. He spoke with no hint of cynicism, of sarcasm, of guile. He only spoke of how happy he was to raise his baby girl, Genesis, here in America and to be able to buy nice clothes for his family.

When he dropped me off, I sat down on my pack and covered my eyes with my hands to hide the tears streaming down my

cheeks. This was neither the first nor the last time I had difficulty bearing other people's generosity. Even though I had liked
to think I was a solo adventurer, I realized that I was never really
alone. I walked a tightwire above a net of compassion, stretched
out by the hands of strangers. My dear countrymen . . .

A week before, when I'd stuck my thumb out on that blue-
skies May afternoon, I had, for the first time, surrendered control. I'd dropped the clock, abandoned the plan, and severed the
puppet strings. It was no longer I, nor my family, nor my local
school board, deciding my destiny. My life was now in the hands
of something else. And now that no one – not even I – was at the
wheel, I felt an odd sense of empowerment. Sometimes, to cede
control to fate, I realized, is to assume more control than ever
before. I wasn't just traveling anymore. I was traveling outside
the formula. I might as well have been floating through space,
trailing my hand in stardust.

I read lefty magazines in Boise, Idaho, almost got run over on
the interstate outside of Salt Lake City, and slept in a Mormon
church in Park City, Utah. In Wyoming, a blonde wearing a cab-
driver's cap asked if I wanted to spend the night at her place. In
Denver, I met up with a friend. Across all the states, I listened to
the purr of rolling wheels atop prairie-flat interstate song lines.
I sped beneath sparkling constellations in a chrome-blue Wyoming sky, alongside a conveyor belt of yellow dashes.

In Nebraska, there were cornfields. My driver, Tom, a
twenty-nine-year-old cook who had been born in South Korea
and adopted by American parents, was on a road trip from Ore-
gon to the East Coast. He picked me up in Colorado and would
take me all the way home to Niagara Falls. Around the time we
drove through Lincoln, Nebraska, it was getting dark, so we
decided to get off the I-80 in hopes of finding a quiet place to
park and set up camp for the night.

On a sleepy country road, we happened upon an abandoned
school next to what looked like a thousand-mile-long forest of
corn illuminated by the moon. A safe place to camp, we thought.

The stars glimmered, and the crickets strung a steady electric hum. We cracked open a few PBRs and boiled ramen noodles on our stoves for dinner.

It was a fairly idyllic scene until we were interrupted by a figure in the dark who advanced toward us with a flashlight. We called out, "Hi there!" but he turned off his light and shuffled away in the opposite direction. Tom and I looked at each other, shrugged our shoulders, and continued enjoying our meals. Fifteen minutes later, another man, also bearing a flashlight, emerged from the cornfield.

"Keep your hands where I can see them!" he screeched. "Put them on the hood of your van!"

Tom and I stood beside each other with our palms flat against the hood of his massive white Chevy. I'd seen scenes like this played out in movies hundreds of times, so I spread my legs as wide as I could, as if struggling to do a split.

"How many of you are there?" the man growled.

"Just us," Tom said casually.

"How many of you are there?" he repeated. He came close enough so that we could see he was a cop.

"Two."

"Is there anybody else?"

"No, just us," we both said, each now with voices quavering.

"What are you guys doing?" he said, oscillating his light from side to side.

"Camping, I guess," I said.

"Do you have anything . . . you know . . . illegal?"

"No."

"Can I see your IDs?"

"My wallet is inside in the center console in the van," Tom said.

"Can I go in and search your van?"

"Yeah, sure, I don't care," Tom said.

"YES OR NO!?" the cop roared.

"Sure, I don't care. I mean, yes!" Tom sputtered nervously.

The cop scoured the van for anything, you know, illegal. I

was still pressed against the hood, my groin beginning to feel the strain. After he was done looking in the van, he dipped his hand deep into my back pocket, giving my rear what felt like a deep-tissue massage as he struggled to fish out my wallet.

"This is private property," the cop said. "You can't be here. Why do you think it's okay to be here?"

"We didn't know," I said. "Sorry."

"Well, you can't be here. People get pretty heated in these parts. You guys just wait here. I'm going to see if anything was stolen." He marched over to the abandoned school that, from what we could tell, only contained a scattering of wooden chairs draped in spiderwebs.

As I watched the cop wave his flashlight back and forth in the school, I thought there was something strangely tragic about the scene. I can't say when it happened, but at some point in the last forty years, it seemed as if something like a giant crop duster flew over our once wild and free country, sowing fear into the belly of America. People now are afraid to take walks at night. Parents won't let their kids wander through the woods. And young men and women are reluctant to hitchhike or embark on adventures for fear of all the terribleness out there that we've been daily reminded of.

The cop came back from the abandoned school clearly more at ease. He told us the name of a campground where we could spend the night. Before we went our ways, he said, "You know, you shouldn't be hitchhiking in this day and age. Times are different now." He looked wistfully at the cornfields, perhaps imagining a different time – some time long ago when the world was supposedly a safer, kinder, nicer place.

"It ain't how it used to be," he said.

On my hitchhike, I'd seen a different country, a different sort of people.

"I know, sir. Thank you," I said, smiling to myself, like I knew something he didn't.

9

VOYAGEUR

July-August 2007—Ontario, Canada
DEBT: $16,000

AFTER FOURTEEN DAYS OF HITCHHIKING, I made it to my boyhood home in Wheatfield, New York. I spent a few days reconnecting with friends and sorting through all the voyageur gear I'd purchased from online reenactment stores with my leftover tip money. (Long story short: My mother was *not* pleased with how I had decided to travel home.) From there, my parents drove me five hours to Ottawa, Ontario, where Bob lived and where we'd set off on our voyage.

Because I'd worked almost a full year in Coldfoot and was ahead of schedule on my debt payments, I had a few months before I had to worry about finding another job. And because I'd almost cut the debt in half and realized that I'd been blowing the debt's oppressiveness out of proportion, I no longer saw the it as some mortal foe worthy of the attention I'd been giving it. The debt was more like some distant cousin I didn't like but still had to interact with at periodic family gatherings. It was a canker sore on the roof of my mouth or effluent from a nearby chemical factory: the debt was something I was

constantly aware of but was nothing more than a mild distur-
bance. Between my delightful spring in Coldfoot, the hitchhike,
and now this voyage, it seemed that I really could *live* when in
debt. And it seemed that I might have found some sweet spot
between fulfilling financial obligations and satisfying adventur
ous longings.

We would be a crew of five. Together, we'd paddle for two
months and 1,500 kilometers along the rivers and across the
lakes of Ontario, Canada, taking a route similar to the one that
Samuel de Champlain had explored four centuries before. After
descending south down the Rideau Canal, we'd skirt west along
the northern rim of Lake Ontario, paddle up the Trent-Severn
Waterway, and zigzag around the islands in Georgian Bay,
which is the watery goiter bubbling out the east side of Lake
Huron. From there, we'd paddle east along the French River,
across Lake Nipissing to the Mattawa River, and down the
Ottawa River to complete a full loop back to Ottawa.

The voyageurs would follow a similar route on summerlong
expeditions, hauling supplies in birchbark canoes to western
forts where they'd collect fur pelts to bring back to the cities
in the east. The voyageurs were a rugged, raggedy, reveling fra-
ternity of outdoorsmen renowned for their impertinence, their
bravado, and their unequaled tolerance for the rigors of the
wild. They lived a grueling life, paddling and portaging from
dawn to dusk, but finding comfort in the songs and rum and
camaraderie of working alongside men a safe distance from the
restrictions of village life back home.

The five of us would live like the voyageurs. All our clothes
and gear were "period correct," meaning that they were made
in the same style and with the same materials of the eighteenth
century. We'd start our fires with flint and steel, sleep under
wool blankets and cotton tarps, and live without toilet paper,
bug spray, water filters, and backpacking stoves. We'd travel by
paddling two very leaky and expensive birchbark canoes that
had been built by a master craftsman in Quebec.

Of course, we could never live exactly as the voyageurs had: we'd be paddling next to houseboats; we'd have to use canal locks; and in the towns we passed through, because we hadn't yet become impertinent and full of bravado, we'd be unwilling to drop trou in front of picnickers, instead favoring flush toilets at canal restrooms. We were forced by law to wear floatation devices around our waists that could be inflated by the tug of a cord. We also carried a GPS, a cell phone, and a camera to document our expedition.

Frankly, I didn't know what the hell I was doing. *This was all just so weird.* Yet I knew the experience would be memorable. And I hoped the strain of the voyage might somehow fast-forward my development. The more miserable the voyage was, I figured, the more quickly I'd grow, as I had after my Blue Cloud climb. But this was no short daylong hike like I was used to in Alaska; this was two whole months of tireless physical exertion — a full-fledged adventure that I hoped would push my limits. Maybe I'd be able to do something gallant and manly? I had ludicrous fantasies of warding off black bears with my paddle, delivering momentous speeches when the crew's morale was low, or saving a fellow voyageur in boiling rapids, bellowing, "*Grab my hand!*" before clasping forearms.

My parents and I arrived in Ottawa and spent the night in a hotel. In the morning I was supposed to show up at Bob's home in my voyageur clothes, but I was too embarrassed to walk through the hotel lobby in my breeches, knee-high socks, and baggy cotton shirt, so I decided that I'd put on my clothes in the backseat of my parents' car in the parking garage.

As I tried to roll my navy-blue breeches over my hips, I realized that I'd made an incalculable blunder. The pants that I'd ordered from an online reenactment store — I was horrified to realize — didn't fit. (Why I hadn't bothered to try them on beforehand is beyond me.) I scoured the rest of my gear and realized, again to my horror, that I'd accidentally left my other pair of breeches at home. With a final tug of the pants, one of the buttons rocketed off and pinged against the ceiling. My mother was waiting outside the SUV, so I handed her the breeches, button, and my sewing kit, and asked, close to tears, "Mom, can you please sew this up for me?"

When we arrived at Bob's, I met the rest of the four-man, one-woman crew.

There were Christian and Diane, both experienced voyageurs who, like Bob, had been members of the original *Destination Nor'Ouest* TV series, which made them minor celebrities in French-speaking Canada. Christian, thirty-two, was a cocky and crass Métis (half white, half Native American) who made a living doing voyageur presentations at elementary schools. One of the first things he said to me was "Looks like you have a little pubic hair growin'," referring to the stubble on my cheeks. "If you smile just right, it looks like a pussy." Diane was a fifty-one-year-old Quebecoise who spoke just a tiny bit of English. She was an expert cook and baker, skilled at finding wild greens to complement our diet of salt pork, pea soup, and bannock.

Then there was Jay, a tall, scarecrow-esque fifty-four-year-old retired substitute French teacher who was obsessed with the history of the voyageurs but, like me, had never been on an expedition of any sort before.

And finally, our leader, Bob, a wealthy fifty-four-year-old motivational speaker and self-proclaimed "World's Foremost Voyageur." Atop his cleanly shaven head rested a felt hat – Indiana Jones–style – that shadowed his long gray beard. He wore a loose hemp shirt and a pair of full-length charcoal denim pants, and around his waist was a frayed red sash: the customary garb of any self-respecting voyageur. Bob had planned and paid for the voyage.

We spent the next hour at Bob's organizing gear and getting our first lesson on packing bedrolls. A bedroll is the finished product of all your gear properly "rolled" and ready for canoe travel that also functions as your seat in the canoe. In my bedroll, I carried an extra shirt, a toque (something that looks like a Santa Claus hat), a capote (a thick, heavy wool coat), an extra pair of cotton socks, an extra pair of moccasins, two wool blankets, and a ground tarp that I'd sleep on every night. The bedroll was tightened with a leather tumpline that is strapped around your forehead, allowing you to easily carry all your gear when portaging over the trails we'd take to bypass rapids.

The rest of my gear included a wooden canteen and a white cotton bag the size of a large purse that held my sewing and fishing kits; knife; flint and steel for starting fires; tin cup for eating, drinking, and bailing; small frying pan for baking bannock; spoon made of bone; lye soap; ropes; journal; and brass pencil. Communally, we shared a seventy-five-pound bag of salt pork and two fifty-pound bags of flour and peas. We had salt, pepper, black tea, bars of maple sugar, bags of dried cranberries, an ax, a large pot, and canoe repair supplies: extra slabs of birch bark and a marine pitch to seal holes.

When Bob asked if I could help him move one of our two ponderous birchbark canoes, I bent over and felt the seams of my breeches strain. I paused. I had been here before. Another ounce of pressure and the seat of my pants would split open like a high school football team running through a paper banner.

"Bob, I hope I lose weight on the voyage," I said sincerely. "I'm not quite fitting into my pants."

He gave me a once-over, leaned over with a good-natured smile, and said, "Ha! You got them on backward."

The pants episode was how the first couple of weeks went. The crew, before they'd met me, presumed that I'd be an able seaman and a competent outdoorsman (given my rafting experiences in Alaska), so they were surprised to learn that they were voyaging with someone who'd never even been in a canoe before. And, of course, I didn't have to make a confession; my ineptitude became obvious when it was my turn to act as the rudder in the canoe's rear, sending the canoe into frenzied figure eights. I also burned the pea soup, filleted my fingers when I tried to get a spark by smacking the flint and steel together, and pretended like I knew how to tie more than one knot, improvising elaborate creations that no one could either mimic or untie.

I may have been inept as a voyageur, but no one had ever met so eager and so willing a student.

Bob focused on helping me fine-tune my paddle stroke. Watching me from the stern of the canoe, he'd say, "Reach for more water. Less arm, more torso. Keep the paddle parallel with the boat. Sink it all the way down. Stop at the hip. Slice the air. Hold your arms straight. Faster. We need forty-five strokes a minute!"

Christian — harassing me the whole time — taught me to sling up the tarps at 45-degree angles with our paddles, a half dozen important knots, and how to patch up and repair the canoes, which were so leaky they needed to be bailed out every hour and repaired each night.

Diane, with far gentler methods, taught me how to bake the bannock with maple sugar and cranberries mixed in and make dumplings from wild apples and tea from cedar leaves.

I tried to compensate for my many blunders by hauling as much gear as I could on portages and assuming all baking and pot-washing duties. And while I couldn't read the weather or identify poison ivy (which I'd accidentally wiped myself with one day), I found that I could paddle just as long and as hard as the rest of them.

After my first week, I tore a page out of my journal and wrote a letter to Josh.

Hey buddy,

This is my typical day as a voyageur:

4:30 A.M. — Wake up. Portage gear from camp to dock with a leather tumpline strapped around my forehead taxing nonexistent neck muscles.

5:00 A.M. — Begin paddling.

7:00 A.M. — Stop for pork and pea soup breakfast that was cooked the previous evening.

7:20 A.M. — Continue paddling.

3–4:00 P.M. — Stop paddling.

4:01 P.M. – Roll out gear, usually soaking wet. Locate fire-wood, begin boiling salt pork and peas. Bake bannock.

5–7 P.M. – Personal time: give attention to aches and wounds, sew up torn clothing, write in journal, etc.

7 P.M. – Eat pork and pea soup. Sit by bonfire; drink ration of rum to dull senses for sleep.

8 P.M.–4:30 A.M. – Struggle to fall asleep as every mosquito in the world tries to sneak under my wool blanket; worry that ants will spelunk into orifices like cave-divers; think about suffocating Bob and Jay in their sleep because they snore like elephants.

Despite the hardships, I'm happy to be doing this. It's a test and I've never been pushed this hard. There is little-to-no "enjoyment," but I feel that today's sacrifice will be tomorrow's reward.

Hope you're doing well with the debt.

Later dude,

Ken

Life was good and simple. We woke up, paddled, cooked, sat by a fire, and went to bed. I had one tin cup that I would use for tea, for soup, and as a latrine when we were too far from shore.

I shat in the woods, wiped myself with leaves, and bathed in the river. I used sand for cleaning pots and twigs for cleaning my teeth. I drank straight from lakes and rivers, and I stopped caring about my smell and unsightly shoulder hairs and my physical appearance in general. We saw the sun rise and set every day. We sat by the fire and talked, or they talked and I listened.

When on the river, I couldn't wait to get off. Yet by morning, I couldn't wait to get back on, lured by the cool morning air that always smelled of sweet black tea. Or the soft gulp of the paddle plunging into water that would send two tiny water-twisters spinning behind each stroke. Or a pair of blue herons gracefully walking along the riverbank on stilts. Or the loon's yodel. Sometimes, we'd paddle in silence through a blizzard of lime-green moths, feeling their soft flaps and velvet tickles.

While we were on the Rideau Canal and Trent-Severn

Waterway, we often floated through towns and alongside con-
voys of huge houseboats that were manned by portly, shirtless
"captains." But when we paddled into Georgian Bay of Lake
Huron, we felt for the first time like we'd traveled back to the
eighteenth century, wandering into manless, boatless, wakeless
waters. We curved our way around thousands of small granite
islands that rose up out of the water, plump and gray, as if some
stone giant was flexing his muscles beneath. Here, the sea and
sky were so blue and calm that it was difficult to tell one from
the other. Bob and Christian, in the canoe ahead, looked as if
they were paddling atop the wavy drafts of the stratosphere.

At night, we set up camp on one of these islands. I went to
the opposite side, took off all my clothes, and went for a swim.
I stood up on a large boulder, letting the wind pant against my
naked body, taking in a sky that had been dyed purple by the
setting sun, which, just below the horizon, projected a mur-
derous bloodred spotlight into the clouds. I had an odd sense
that I had been here, in this very place, long ago; that I'd experienced these very sensations in a different lifetime. I imagined
an ancestor of mine, perhaps in this very spot, or in a spot like
it, a thousand, twenty thousand, or a hundred thousand years
before, thinking about how an ancestor of his might have been
thinking these very thoughts.

I found myself daily shedding the traces of my century, leaving the comforts and conveniences and distractions of it in the
wake of our canoes. After a month of paddling thirty-two kilometers a day, my arm and shoulder muscles had become taut
and sinewy; my hair was greased to the point of waterproof-
ness; my face, begrimed and sun-baked, took on a swarthy complexion; and on my chin and cheeks sprouted a bushy brown
beard.

The relentless toil of voyageur life made me indifferent,
even contemptuous, of what now seemed to be the frivolities of
civilized life. Convention, decorum, proper demeanor – these
things were useless out on the water. They wouldn't help me
get through nights when squadrons of mosquitoes would buzz

in my ears and creep under my blankets. Nor would they make the pains in my feet go away after mile-long canoe portages over outcroppings of sharp rocks in my thinly soled moccasins. Nor did they soothe my shoulders after twelve hours of almost nonstop paddling. When your life is all toil and hardship, the things that matter and the bullshit that doesn't become easy to separate.

I started to feel less like a blundering suburbanite and more like someone who kinda knew what he was doing. I could tie an assortment of knots and I could start fires with my flint and steel on one swipe. I cooked all our meals, washed all our dishes, and never complained. I was even having some luck predicting storms by observing cloud movements, moisture, and the wind.

Christian, who was just starting his career as a traveling speaker, only joined us for about half of the voyage. He and I, toward the end of his stay, became canoe partners. He was flagrantly crass, singing French songs about taking a dump in -38°F weather, farting obnoxiously with the tilt of a hip, and once, when he drank a bottle of Madeira, showing his "specialness" by pulling down his pants and tucking his testicles into his closed thighs.

But one day, on the Mattawa River, he forgot to maintain his façade of crassness and self-absorption, unveiling to me just how spiritually sincere he was. In hushed tones he confided that he has a "gift." During moments of deep meditation, he told me that he's able to see into people's minds "as clearly as the horizon before us" and that, in certain settings, he'd even had the power to view people's dreams. He told me how his native ancestors would go on "vision quests," which was a ritual that young men in villages would undergo when they reached a certain age. They'd venture out into the woods and starve themselves for days on end until they were granted a vision. What they saw during their quest would play a huge role in the formation of their identity. Sometimes they'd even change their name according to whatever animal spirit greeted them. It sounded a bit fantastical, yet I totally got it. The vision quest was like a journey. When we are raised by institutions, we are fashioned, in ways big and small, to be like everyone else. But when we go on a journey – especially a journey that follows no one else's footsteps – it has the capacity to help a person become something unique, an individual.

While Western society never had anything quite like the vision quest, we do have a heritage of journeying laced into our cultural DNA. In the 1930s, Americans hopped trains. In the 1950s, beat poets wrote about road trips. In the 1960s, we hitched rides. Today, however, it seems like the whole "coming of age" adventure has been abridged from a young person's life experience, leaving no gap, no bridge, no moment of real freedom in between school and career.

I listened carefully to what Christian had to say of dreams and visions and seeing into people. We'd been paddling for twelve hours straight. It was scorching hot, and for whatever reason I couldn't think about stomaching another bowl of pea soup that day, which we'd been eating for breakfast, lunch, and dinner for almost six weeks. Toward the end of the trip, I became transfixed with my shadow to the left of me on the river. My body was colored in with a deep, dark blue and striped with the water's spiny ridges. I watched the shadows of my arms and paddle moving rhythmically, hypnotically circling my body with each stroke. I thought I could see the eyes of my water-shadow. A scattering of water droplets flew from the paddle's tip as I brought it forward, the blade lightly, with a surgeon's precision, skimming the water's crest. It was as if someone had removed my eyes, then jammed two new balls into the sockets. While I wasn't having a vision, I was feeling the purifying glow of the ascetic act, experienced when – in the midst of a hurricane of struggle and strain – you come upon the "eye" of your adversity, when, for just a fleeting moment, the storm breaks and you're afforded a flash of something sublime.

Squirrels ran over my blankets at night. I slept in rain, in blustering winds, and with mosquitoes buzzing in my ears. Underneath each of my big toenails were dark bruises. The skin on my feet was dry and cracking. On portages, we'd have to carry the canoes, which would grind into our shoulders, making it feel like our muscles were being flipped off the bone. The tendons in my arms felt like they were going to snap like rubber bands, and sometimes I'd have back pains that made it hard to breathe. Yet all variations of pain and strain turned to numbness. No longer a man, I was a paddling machine. I felt like there was no discomfort that could be too discomforting. There was no day too long or load too heavy. I didn't miss toilets, or showers, or homes, or cars, or anything modern. For two whole months, I could carry just about everything I needed in the canoe or on my back, with relative ease, making a whole houseful of needless stuff seem silly.

My relationship with nature was changing. No longer did I think of it as something to conquer, like a mountain summit. Nor was nature something to be glorified, which we tend to do at scenic road pull-offs. Nature, to me, was no longer beautiful. Nature, I realized, is only beautiful when you're at a safe distance from it. Watching a setting sun from a windshield can mean romance, serenity, beauty. On the water, though, it was a warning for mosquitoes, storms, and the cold. When I was mesmerized by nature before, I was merely disconnected from it. After more than forty days on the voyage, I no longer saw nature and myself as independent entities; rather, I was nature, living among the roots, insects, animals, and storms. Because nature was indifferent to me, I began to feel indifferent to it.

The voyage was teaching me how unexceptional I was and how exceptional the human mind and body is. What wonders the human mind and body are capable of achieving! How so few know how much we can do! Our limits are merely mirages on the far side of the lake—we can see them ahead, but that's all they are: mirages. Our real limits are beyond the scope of our vision, beyond the horizon, a boundary worthy of our exploration.

Just as I noticed that I was changing, I saw that Bob was

changing, too. Diane and Christian had left the voyage to go back to work, so they were replaced with an affable, though slightly out of shape, guy named Art, whom Bob had met and recruited at one of his speeches. Now, it was just Jay and me in one canoe, and Bob and Art in the other.

At this point Bob ceased consulting the group about decisions. He became short-fused and barked orders at me while picking on and bullying Jay, swearing at him for minor blunders. He went from being a stern, though reasonable, leader-elect to a hot-tempered monarch.

"Listen—when I say something, you guys listen!" he screamed at us. "We can talk about who's right and who's wrong later!"

He began shouting orders about minutiae and easy tasks, like lining the boats around some swift-moving water. I just wanted to scream, "Bob—shut the fuck up! We got it!" But never did. This went on for weeks.

My anger was all-consuming: anger that I directed as much at Bob (for bullying me) as at myself (for doing nothing about it). Instead of feeling like an intrepid voyageur, I felt more like a lowly deckhand, assigned the nightly task of emptying his master's chamber pot. If one of those eighteenth-century landscape painters had drawn a portrait of our crew, I'd be one of the surly laborers in the background, wearing tattered rags, hunched over from all the carrying, cooking, and paddling, while Bob—the courageous leader—would be posed erect, with a foot on the canoe's gunwale, a look of gallant determination on his face, and a generously sized bulge in his breeches.

Each time he raised his voice at me, I didn't think I'd be able to handle one more command, one more insult, one more assertion of his superiority. I relied on my usual tactic of bottling my anger and hoping the problem would blow over—a tactic that had never worked in the past, yet was one that I continued to apply to situations with an irrational faith. And then the rapids incident occurred.

On the French River, Jay's and my canoe was sucked into

some swift water. Taking a birchbark canoe into swift water was a big no-no because the brittle hulls couldn't handle the slightest bump against a rock, which would be all the more difficult to avoid in swift water. Bob, from the shore, watched as we got plunged downriver. Jay and I, with ease, navigated around rocks, but that didn't stop Bob from unloading on me. He screamed and yelled and cursed, selecting his words from a vast array of uninterpretable French expletives. He went on for some time, but I only remember one line in English: "What the fuck did I say?! Real smart move!"

I couldn't take it anymore. I couldn't think about anything but confronting Bob. I stared at him ahead, hoping that by sheer exertion of will I could make his skull combust into a cloud of pink mist. My blood pulsed to the rapid drumbeat of rebellion. I wanted mutiny.

I decided I'd march up to him when he was alone and grab his neck. The sky would blacken, flames would blaze behind me, and I'd command him to never swear at me again.

Bob had started putting these voyages together on his own largely to boost his national profile for future speaking contracts, but also to pay homage to his ancestors, many of whom were voyageurs in the seventeenth and eighteenth centuries.

The voyageurs back then came entirely from the lower classes. They could have lived reasonably free lives as peasant farmers, but they chose to be voyageurs. They sometimes worked fourteen hours a day, paddling at a rate of forty-five strokes a minute. On portages several miles long, they carried hundreds of pounds of gear using tumplines strapped around their foreheads. Many voyageurs didn't live past their forties, dying of crippled backs and strangulated hernias.

They chose that life. They favored the winding river over the cobbled road, strain over comfort, adventure over monotony, full lives over long ones. "Voyageur" means "traveler." They traveled for a living.

What was I traveling for? What was anyone traveling for? In

centuries past, there was always some blank spot on the map that needed to be filled in, terrain that needed to be discovered, goods that needed be shipped across perilous oceans. The voyageurs traded goods, supplied forts, and got paid for it. These people all had very clear reasons for going on journeys. Yet here we were, pretending to be people from a different century.

The twenty-first-century adventurer – because he has no frontiers to settle or wild lands to explore (nor the technology to push the boundaries of outer space) – has to sort of make it up. And that's what the four of us were doing: creating a journey out of nothing, going back two centuries so we could feel what the twenty-first century wouldn't let us feel.

Yet, despite the inauthentic nature of our voyage, there was something undeniably "real" about it. Having spent a year in the arctic and a summer on the water, I was presented with a new vantage point from which I could finally see civilization and suburbia for what they were.

In suburbia, except for when we conjure the willpower to go for a walk around the neighborhood, there's hardly any real purpose in going outdoors. There are no fences to repair, no bean fields to hoe, no water to fetch from the stream. Back at my parents' home, whenever I was hungry, I'd grab food from the well-stocked pantry or fridge. My water would come from magical sinks and my heat from magical vents. Because I was not really needed for anything, I'd spend my time fulfilling desires: watching TV in the family room, reading books in my bedroom, and playing video games on the computer.

On this voyage, I couldn't help but think that we *need* need. We need to be forced to go outside. We need to be forced to depend on one another. We need to be forced to sacrifice, to grow a garden, to fix a roof, to interact with neighbors. Nature had been all around me as a boy. It unleashed terrifying storms, spun circular cycles, inflicted bone-chilling cold, and renewed itself with springy revivifications. Yet I was completely oblivious to it all. I was playing video games.

Even though I'd become livid, frustrated, and demoralized

for being sworn at and ordered around and for not being treated like a grown man on this voyage, I was glad to have back the sensations my century had deprived me of, lust for mutiny and all.

That evening we set up camp on a large, forested island next to the French River. Bob was cutting up the salt pork: his evening ritual. I went up to him, as planned, ready to grab his neck and deliver my jeremiad.

I couldn't, though. I just stood looking at him, paralyzed by my timid nature that had been holding me back all these years.

We all went to opposite corners of the island to lay out our tarps and wool blankets that we'd sleep on. As I laid out my ground cloth and began erecting my tarp, I thought about how, over the past six weeks, I'd put myself through a never-ending gauntlet of torture and pain but was still standing, still holding up. And I was paying off my own debt, going on my own travels, and living a free and independent — albeit poor — life. I wasn't sure why it had taken me so long to realize it, but I knew I was no longer the sort of person who should let someone else kick me around.

I left my stuff and found Bob's spot, where he was getting ready to bed down on a smooth boulder. I walked over sheepishly, head down, going over what I'd say one last time.

I started off weakly. "Bob, do you have a minute? About earlier at the rapids . . . I wanted to say that it was completely my fault."

"Oh, well, that's all right," he said conciliatorily.

"But, Bob," I said, looking into his eyes, "I never want to be yelled at or sworn at again."

We traded a few more words and shook hands. He said, "Good man."

10

....

CORPSMEMBER

October 2007-March 2008
Gulfport, Mississippi
DEBT: $16,000

AFTER JOSH SPENT THE SUMMER as a tour guide in
Coldfoot, he moved back in with our friend in Denver.
His job search hadn't worked out the last time he was in Denver,
but he still held out hope that he'd find a long-term job there.
While the seasonal work afforded him some freedom – to move
around, to meet new people, and to work for different employers – the transient life was also extremely inconvenient because
the gaps in employment had made it difficult for Josh to pay off
his debt at a reasonable pace.

Denver, when he arrived, proved as inauspicious as ever.
After several more rejected job applications, he enrolled in
bartending school. It was a decision he made not only because
he'd lost hope of finding a job relevant to his education, but
also because he'd always harbored fantasies of one day becoming some cool and charming Sam Malone–type bartender,
telling lewd jokes to a trio of portly regulars, swiping a towel
across a polished bar, and flashing a boyish smile at one of the

waitresses. But like almost all of his ideas to date, this one bombed, too.

To: Ken Ilgunas
From: Josh Pruyn
Date: October 12, 2007
Subject: fuck Denver

Where to begin . . . I guess I should start by saying I've applied to approximately 15 bartending jobs, but I couldnt even sell myself for a position a high schooler could nab. I had a 3.83 gpa, a degree with a double major, numerous academic awards, have completely open availability, good references etc., etc., etc., and I can't get a job that pays under minimum wage. Holy god is this frustrating . . . all the education I have received at this point (college, graduate and bartending) and all the hard work and effort I've put into all three, and within a week I'm going to have to resort to being a waiter – a job someone that never went to school could get? Plus, I have $66,000 in debt.

Running out of options, Josh applied to be a waiter at a Red Lobster and as a clerk at a fitness center. But they didn't want him, either.

Finally – finally! – a friend of a friend helped him get a job at a for-profit online school called Westwood College (which also has seventeen trade schools nationwide). It paid $16 an hour and it offered the standard benefits like medical, dental, 401(k), and sick days. Josh's worries – at least for now – were over. It was an occasion for relief, no doubt, but his initial reaction was ambivalence. "Of course I hope I enjoy it," he wrote, "but I also fear that if I do, I will fall into the ranks of men who compromise their dreams for comfort and security."

Josh had officially entered Career World.

Meanwhile, I'd been struggling to find a job of my own. After the voyage, I spent a month at my parents' home filling out job applications so I could re-declare war on my debt after our summerlong cease-fire. I wrote a couple of freelance articles for an alternative weekly newspaper in Buffalo that I had interned with in college, but after sixty hours of writing, and $120 for my services, it was clear to me that I'd have to find some other way to make money.

I found temporary work in Gulfport, Mississippi, as a corps member on an AmeriCorps trail crew called the Gulf Coast Conservation Corps (GCCC). For two and a half months, I'd get paid $250 a week to blaze trails, remove invasive species, plant trees, and clean up the mess Hurricane Katrina left two years earlier. While the pay was meager, room and board were provided, plus I could expect a $1,000 "education award" at the end of my term that could be put toward my loan.

Twenty of us lived in two gender-designated barracks that had been set up between two Little League baseball diamonds in the middle of a gang-ridden and mostly black ghetto in the city of Gulfport. The baseball diamonds and barracks were neat and austere, but the surrounding community was graffitied with a third-world grime and littered with a war zone's squalor. Homes were falling over, lawns were covered with garbage, and ditches were brimming with beer cans and bottles of malt liquor. I'd figured the place was going to be a mess because of Katrina, but it was evident that the town had been in ruins long before the hurricane swept through. Upon viewing the devastation, I could hardly believe I was still in America.

When I met my fellow crew members, I felt like a camp counselor wrongly assigned to the role of camper. Except for the four team leaders, I was – at the age of twenty-four and with my bachelor's degree – the oldest and most educated on the half white, half black crew. The rest of the crew was mostly in their late teens. Many had yet to receive their GEDs, several had kids whom they'd abandoned or were raising on their own, and others had dealt with or were dealing with alcoholism, drug abuse, and depression.

Lyle, eighteen, was a 350-pound white kid from the swamps who, despite his massive size, had been bullied all his life. He'd never kissed a girl or driven a car, and he didn't even know how to use a washing machine.

Grant, twenty-one, from a well-to-do family in Maine, was souped up on antidepressants that his father, also his psychiatrist, had prescribed to him.

Owen, twenty-two, thin and gangly, had been addicted to meth and had already done time. He was so poorly educated he didn't even know who our president was.

Robert, nineteen, was a lean, vivacious black kid who had eighteen brothers and sisters, plus a child of his own.

Jacey, nineteen, was studying to get her GED and would be one of the many girls who'd get impregnated while at camp. She'd call Mclinda, one of the crew's single moms, a "muthafuckin' bitch-ass hoe," which – with the right tone – can be, and in this case was, a term of endearment.

I was the crew's ascetic. After Coldfoot and the voyage, I was accustomed to an austere, rigorous, nearly possessionless lifestyle. I brought nothing with me except three sets of clothes, my tent, a sleeping bag, and a few books.

After we got back to camp from the trails, I'd do push-ups and run laps around the baseball fence. Most everyone else on the team chain-smoked cigarettes, smoked pot when the leaders weren't looking, and squandered their paychecks on alcohol. At night, a new combination of corps members would get together in the concession stand to fornicate.

I fantasized about being back at college and around students who wanted to change the world and blaze happy, healthy futures for themselves. If I didn't get out of Mississippi before long, I worried I'd be swallowed whole by my surrounding culture: I'd buy some property in the bayou, put up nineteen NO TRESPASSING signs, draw out one-syllable curse words into raspy haikus, and inseminate everything that moved. For too long I'd felt like a seed blowing through a desert. Unlike Coldfoot or Gulfport, a leafy green campus seemed like good soil in which I could plant myself and grow.

• • •

On weekends, I'd fill up my backpack with camping gear and walk across a log that had fallen over the murky brown Turkey Creek. There, I'd set up my tent on a bed of pine needles, surrounded by mammoth live oak trees that bore hundreds of thick, snarled, muscled arms, and listen to the creek gurgle onward to the Gulf.

Even in the winter months, Mississippi teems with life, located as it is along sweltering southern latitudes and beneath a nonstop succession of ocean-borne storms. One cannot leave the Gulf Coast region without an impression of the terrifying fecundity of the place or the unrestrained concupiscence of its inhabitants.

When in a Mississippi jungle, you feel as if you're at the mercy of dark desires and ancient impulses. Despite unprecedented levels of pollution, cancerous suburban sprawl, and devastating natural disasters, the animals and insects still thrive in Mississippi. But not as much as the humans, the worst of all in Mississippi's animal kingdom, who reproduce with as little forethought as the cicadas restlessly moaning for mates in the bayou.

Mississippi — with its sister Gulf Coast states not far behind — has the highest teen birth rate in the nation. Hammered onto pine trees on even the most desolate backwoods roads are signs advertising reduced-rate DNA tests. In the little bit of time I served with the GCCC, among the thirty members who came and went from our crew, eleven of them — yes, eleven! — either got pregnant or impregnated someone else.

In Mississippi, for the first time in my life, I'd become an object of attraction. Not only did one of the single moms start to court me, but a gaggle of Mississippi girls would lavish my "bootie" with compliments when I rounded home plate on my jogs. Owen told me his girlfriend's girlfriend wanted to "get with me," and while I was no doubt flattered, I politely declined all invitations, partly because I recognized that all the males were getting this sort of attention.

In Mississippi, everybody's getting laid.

Because I'd been more or less single and celibate for years, I could identify with those sturdy, southern live oak trees that were among the few that Katrina had failed to yank from the earth. I knew what it was like to be alone – a forest of one – forever holding fast to the ground amid tempests of temptation.

I was again dedicated to my goal of paying off the debt, and I couldn't let the lure of material items, alcohol, or girls cause me to topple, however tempting they were. I left the male barracks, took my tent out to left field, and slept in it every night to put space between the hordes and me.

Despite the low pay, I loved being out on the trail, wielding a pick-mattock, Pulaski, or ax for hours on end, carving lanes through logs that had fallen onto trails, delighting in the strain of muscles, my torso and arms lubricated by the quarts of sweat that flooded out of my pores, lumpy droplets trickling down my back like a procession of ants. In the midst of good, steady work, all conscious thought comes to a halt. Neuroses vanish as if they never existed. I was engaged in a constant state of sensory distractedness: the arc of the ax, the flying chips of the log, the sweet fragrance of minty pine needles. With distraction would come peace; during those moments, there was no more debt, no more aching desires, and work no longer felt like work but a happy engagement of mind and body that could be rightfully confused with the joy of all joys, the epitome of human existence.

We submerged ourselves nipple-high in swamps to drag out tires, cleaned the beer bottles out of Gulfport ditches, and walked through pine forests, pulling out rafters, appliances, and anything else Katrina had tossed there. Normally, I was engrossed with the work, though sometimes, in the jungle, I'd find myself distracted upon catching sight of Sami, a tomboyish nineteen-year-old crew member from Minnesota.

Like me, Sami kept mostly to herself. She was quiet and focused and one of our hardest workers. She'd be sawing off a branch that hung over the trail and I would steal a glance at the cherry-colored hair cascading down her back in frenzied swirls. It was always, from want of a shower, frazzled, oily, lustrous, practically alive. *Just how I like it.* She was the very embodiment of all the female charms that so enchanted me: she had a pale snow-white complexion made ruddy by the sun, spattered with sun-browned country-girl freckles. She wasn't manicured or groomed or dolled up; rather she was the dirty-kneed, smooth-muscled woman of the woods, a wood nymph who attracted suitors less with the artifice of makeup or the exaggeration of a Wonderbra, and more with an irresistible fertility, a delectable ripeness, an enamoring comeliness that made the earth tremble beneath me and set my loins aquiver. I wanted nothing more than to set down my ax, put my arm around her soft waist, and draw her against me. Maybe I'd stick my nose in the back of her hair and glory in the curls and smells. Maybe I'd kiss her sun-hot shoulder and she'd turn around and we'd make gentle love on the warm forest floor.

She'd wear the same thing in camp every day: a baggy blue tee and an ill-fitting pair of jeans. Though the getup did nothing to accentuate her feminine features that other girls wouldn't hesitate to flaunt, her approach turned me on far more, for it communicated to me something far sexier than curves and hips: It was her complete disregard for prevailing norms and up-to-date fashion. And that independence of mind and self-assurance thrust me into a state of constant, hopeless, agonizing, "I will kill to have her" desire.

My previous girlfriend was a Baptist girl from Buffalo. I was a sophomore in college, and she, a senior in high school who carried a full-sized eight-pound Bible in her purse that affected the way she walked. She was passionate and devoted. Not for me, of course, but for her god. To her, I was a mere momentary corporeal distraction, a mere blip on the timeline of her eternal soul — that one time she consorted with an unbeliever. How could I compete? God promised an afterlife, forgiveness, salvation. All I could do was turn clockwise on the gymnasium floor with her at her homecoming dance.

One night, I took her out on a date to get milkshakes at a McDonald's. After successfully getting to first base in the parking lot, the sweltering summer heat and my pent-up longings and a sugar high made me take a desperate, fleeting glance at what wonders might exist on second. She, stricken by religious fervor and offended by my advances, began to describe Christ's gruesome crucifixion in vivid detail, bursting into tears halfway through, announcing that my earthly desires and I were obstructing her "walk with God." When I objected that I was doing no such thing (even though I was, in fact, trying to casually nudge God off her path and down the canyon walls), she muttered hopelessly, "You just don't know how it feels to have Jesus inside of you . . ." After she broke up with me, I decided to steer clear of girls and sex and relationships for the remainder of my college life, if just to stay focused on my schoolwork and keep my sanity intact.

Yet now, after meeting Sami, I couldn't help but reevaluate my policy. She was just so different, so unlike many of the girls I knew. She was more than just cherry hair and comely features. There was something complicated about her. Something complex. There was something about her that made me think she was one of the first full-fledged human beings I'd ever met.

She reminded me of Jane Eyre — the type of girl who might have, at an earlier age, clung to notions of chastity and temperance, proudly wearing a tattered governess's dress, unconcerned about what others thought. She had a wide, cheery white smile and big brown eyes. In those eyes, I saw innocence, but I also saw fatigue. I could tell she carried with her some terrible burden from her past that she didn't know how to shrug off.

Sami was my secret crush.

I spent three wonderful months in Mississippi and was prepared to leave to find work elsewhere, but after TJ, a Mississippian and one of the four team leaders, was fired (due in part to having been caught fornicating with a crew member in the shower), they promoted me to his leadership position.

I decided to sign up for another three-month term. Despite the low pay, I was still able to make my loan payments. And just like at Coldfoot, the free room and board were allowing me to save what little money I made. Plus, now that I was going to be a crew leader, I'd get health insurance, an additional $50 a week (boosting my weekly salary up to $300), and an extra $1,000 to pay back my loans. After three months in Mississippi, I'd paid off $2,000 of my debt, with $14,000 to go.

But it wasn't just about the money. I had fallen for Mississippi, and I thought it would be good for me to assume my first-ever leadership position. Because of the newfound self-discipline that the voyage and the hard work on the trails had instilled, I didn't think I'd have any problem obeying the one rule expected of us leaders: Don't sleep with the crew members.

I started to get to know my fellow crew members and did my best to help them out. I knew from my dealings with Bob that to win their respect I'd never order but ask, and that I'd treat them all as equals.

I learned that they weren't as hopeless as I'd once thought. And contrary to my first impressions, they all wanted to improve themselves. They were off the streets, and the hope they felt was palpable.

"All right, Lyle," I said, "first things first." I started with the basics. I taught him how to fold clothes and turn on a washer. While folding one of his shirts, I told him, in earnest, "You're a good-looking dude. You just gotta lose a little weight." I helped Robert with his math homework and drove him to his GED class every Wednesday. When Grant asked me to take him to the hospital to get more pills, I told him about my correspondence with Josh, which had been my therapy, and I suggested that he didn't have to wash his feelings away with a white capsule. I set Owen up with an e-mail account and tried to get him to take his goal seriously about becoming a chef.

Every day, I took someone to the hospital, to get checked for STDs or if I found out that another female crew member was pregnant. I drove them to get groceries, helped them apply to school and fill out job applications, and listened to their problems (which all seemed so much more ponderous than my own trifling, privileged worries). And I loved every second of it. The work was good and honorable, but to psychologically cope with all the horrors around me, I found myself becoming emotionally detached — detached so I could bear other people's misfortunes. I didn't have any time to think about my loneliness, my debt, or my future worries. I became a robot: I worked and ran and read and slept. It seemed as if it wasn't so much the schedule that was ruling my life anymore but a self-imposed, rigid sentence of self-discipline that I was serving.

Every weekend, the crew would try to persuade me to join them at a karaoke bar called the Salty Dawg. But except for the occasional used book online, I was buying nothing, spending nothing, and saving everything. I was no less reluctant when they invited me to join them on their trip to Mardi Gras in New Orleans, but my defenses began to falter when I found out that Sami was going, too. Much to all their surprise, I agreed to go. Seven of us crammed into a small sedan, and because there were too few seats, Sami sat on my lap, which sent me into a helpless state of excitement and nervousness and barely concealed ardor (among other barely concealed issues that arose).

On the ride over, she said she was looking for her next job, and I told her I could probably get her a job cleaning rooms at Coldfoot, adding that I might live up there that summer, too, because I was applying for a job as a backcountry park ranger at the Gates of the Arctic National Park and Preserve, which has a ranger station in Coldfoot. I knew I could set Sami up with a job easily, but I also knew that I had little chance of getting the Park Service job, as I was still, to say the least, far from being a competent outdoorsman. But if I couldn't impress the Park Service with competence, I'd do so with diligence. Every two weeks I called or e-mailed them, restating my interest in the job. If I got it, I knew my debt worries would be over. I entertained the idea of us both working up there, making love on mountaintops, emerging naked from hidden lagoons.

The way our relationship started was, I'll admit, less *Jane Eyre* and more *MTV Spring Break*. (Starting a relationship at Mardi Gras, after all, may be just a notch above starting one in a porn shop.)

It all began when she asked me to go to a Yonder Mountain String Band concert. I dropped $20 on a ticket, and at the concert, on the dance floor during the first song, she leaned against me, and from that moment forth, I was hers and she was mine.

Thereafter, back in Gulfport, Sami would sneak out of the female bunk after everyone went to sleep in order to bed with me in my one-person tent on the baseball diamond. In the morning, she'd wake up before everyone else and head back to the female bunkhouse in order to keep our trysts a secret.

She was all passion – all feeling and warmth and wet kisses. She carried about her an air of effervescence that was so real and unaffected it was intimidating. She wasn't intellectual, or logical, or deliberate. She was all spontaneity and carefreeness and impetuosity; her actions governed entirely by passing fancies and momentary whims. She lived by a code that I wanted to live by but didn't think I'd ever be able to.

I knew it was dangerous to be dating a crew member, but there was no fighting it anymore. My defenses had hoisted a white flag, my chastity belt had fallen to my ankles, my passions were stirred to a fever pitch. I desired her with a disquieting intensity; my deliberate style of living gave way to the impetuous now. Sami was in my every waking thought. At work, the minutes ticked by too slowly. I couldn't wait to get back to our tent, where life was nothing but caresses and copulation, her cool auburn hair dancing on my bare chest. I was a high schooler again, stupidly, foolishly in love.

During our nightly trysts, she'd ask me to tell her about Coldfoot and my hitchhiking adventures. For some reason, she always wanted me to tell the stories. Hers, for some reason, were too hard to tell.

As we got to know each other, I could tell that we were like two different-colored rivers becoming one, each coming from wildly different sources. I came from the concrete dam of student debt and she from the melting glaciers of depression, where her very existence hung in the balance.

Two years before, Sami had been lying in a hospital bed with tubes jammed into her throat. Throughout high school, she had tried to kill herself on five separate occasions.

She'd had a happy childhood and was raised by loving parents in middle-class Minnesota, but for no clear reason, Sami plummeted into a bottomless, inescapable depression. She stopped eating, and when she did eat, she threw it back up. She'd cut her wrist, OD'd on pills, and chugged peroxide. Clueless and without any better idea, her parents and doctor put her on a high dosage of prescription drugs, which numbed her so much that she'd go to wild parties and get wasted just so she could feel something. At one party, she got raped. She was sent to therapy, hospitals, and even college. Nothing worked.

Providence, though, had other things in mind for her. On a trip to the mall to ask for cardboard boxes so she could move out of the college she'd just dropped out of, she joined a crowd that had surrounded a magician who was on tour raising funds for his animal sanctuary down south. Sami, drawn to the tiger and lion cubs that the magician had brought with him, asked him on a whim if she could join the tour. The exchange marked the first time she'd let her spontaneous instincts lead her in a constructive direction.

A week later, she was working at the sanctuary in Oklahoma, where she cared for displaced animals. She went on tour across the country with the magician to help raise funds for the sanctuary. A year later, she found her way to Mississippi, where she joined the trail crew.

Suburbia, work, school: These, for the adventurous at heart, are no more than boxes – boxes too small and confining for souls made to fly. Because Sami had never had the chance to acquaint herself with the adventurous sensations her soul longed to nourish itself with, she'd seen little reason to go through the trouble of living. Comfort and security, it seems, when overprescribed, can be poisons to the soul – an illness that no amount of love can cure, freedom being the only antidote.

When I met her, she was on a "living high." She reminded me of myself when I decided to drive to Alaska and climb Blue Cloud years before. She was willing to do anything, chance anything, and risk anything if it meant she'd get to feel some new sensation that had previously gone unfelt.

"I'm going to hitchhike to Alaska," she said to me one night.

"What?"

"You know, like you did."

"Sami, I don't know if that's a good idea."

"Why not?"

"Well, it was different with me."

"You mean because I'm a girl?"

"Well, something could happen to you."

"But wasn't it you who told me that the world wasn't as bad as everybody said it was?"

She was thinking about Alaska because she'd gotten the job in Coldfoot. I, on the other hand, hadn't heard back from the Gates of the Arctic National Park, so my reverie of us up there together seemed far-fetched, and I wasn't sure what to do about our relationship. While I didn't want our rivers to branch off in different directions, it seemed like separating was the best thing for each of us. She had a job, and I needed to find work. But now this hitchhiking business changed everything.

"Sami, please don't hitchhike. I'd be worried sick," I said.

"Kenny, I've been told my whole life what I should and shouldn't do."

I didn't want her to hitchhike all alone. Despite her brush with depression and death, despite all the poverty and destitution she saw in Mississippi, she still really believed that this was one big, happy world. She didn't know when to feel afraid. She didn't know when someone was hitting on her, or when not to tell a joke. After her last suicide attempt, she'd experienced something like a rebirth, and she was rediscovering everything anew, with a fresh set of eyes. I loved her for the cheerful world she saw, but I knew it would work against her if she were to strike out on her own so soon.

I begged her not to go, but there was some part of me that knew such an adventure might be good for her, as it had been for me. I knew I couldn't, for the sake of her well-being, shove her back into one of her airtight boxes, so I told her that I'd hitchhike with her, not to Alaska, but to New York—where my parents lived—if she promised not to hitchhike alone.

Meanwhile, Josh, over the course of my six months in Mississippi, grew more and more disenchanted with his new job at Westwood College. He had the title of admissions representative, but he was really no more than a "glorified telemarketer," as he described it. He sat in his cubicle all day and called up teenagers to persuade them to attend Westwood. Most of them came from low-income families who—Josh pointed out to me—were a lot easier to persuade. At first, he thought the job was socially beneficial. He figured selling education was better than selling anything else. But when he learned more about Westwood, he found himself in a moral quandary.

Students enrolled at Westwood's online school pay $64,000 to $79,000 for a three-year degree. Because Westwood—like many for-profit colleges—isn't regionally accredited, students can't transfer their credits to conventional four-year colleges. After a little googling, Josh discovered that hundreds of Westwood students couldn't get jobs with their degrees, nor could they pay back their astronomical debts. Plus, Westwood was lending loans to students with ungodly 12 percent interest rates.

It was a cruel irony that Josh, floundering in student debt of his own, was now in the business of getting other young people to go into debt. This was not the time for idealism, though; Josh needed the money, and morally unambiguous jobs were clearly in short supply. So, like a good loan drone, he shuffled out of bed each morning, groaned, put on his collared shirt, and settled into his cubicle.

Our season with the GCCC ended in March. Sami and I said good-bye to the crew. I still didn't have a summer job lined up, and Sami's new job at Coldfoot didn't start till the beginning of May, so we had a whole month to explore America's East Coast.

Hitching rides never seemed so easy. Rarely would we have to wait for more than half an hour. On my Alaskan hitchhike, I was stranded on the side of the road for hours at a time, sometimes a whole day. Sami was the key: She'd hold the sign while I sat on my pack behind her trying to look harmless (or attempting to remain out of view entirely). Her bright, innocent smile was the perfect bait for the predominant demographic of our drivers: lonely, middle-aged males.

One of our first rides was with Terry, a professional driver who takes train conductors to and from their stations. He spoke with a slur, fumbled with the gear stick, and casually remarked that he had stopped at a bar for a cold one on his way home. But I didn't realize that he might have been drunk until I caught him staring at a fridge full of beer in a gas station, eyeing his options long and hard, as if pondering some great moral question. He bought a Coors tallboy and cracked it open halfway to Jacksonville, Florida, where Sami was hankering to see the beach. Terry, like most of our drivers, had a checkered past. His was fraught with prostitutes, crackheads, and alcoholism, as well as a devastating separation from his wife. Before he dropped us off, he shared with us the six words of wisdom that he'd failed to live by: "The truth will set you free" – a phrase that Sami later stitched into her hemp purse.

Rusty, a trucker, would take us all the way from a thruway entrance ramp in Jacksonville to a truck stop in South Carolina.

He told us about his mail-order bride in Ukraine, and he spoke of his many travails, listing his tragedies nonchalantly, like a mechanic smugly specifying repairs needed after an inspection. It all started when a tornado blew down his boyhood home. Later, burglars stole his inheritance that was kept in a shoebox. His stepfather raped him as a child and, decades later, would have an affair with his wife. Rusty said his wife tried to kill him with rat poison after she had a nightmare – or a premonition – that was so terrifying she miscarried their unborn son. He had had four types of terminal cancer, validating the premonition, and finally, after having a vision on his deathbed, he overcame his illness, converted to Christianity, and became a new man.

Harry, our next driver, said things like "Hey, man . . ." in that nasally, slightly annoying, hippie sort of way. He was forty-four, but he looked sixty-four. He had a long, feral beard and a white ponytail. He had a bad limp, which he picked up a couple of years back when he was sandwiched between a pair of forklifts when he'd been drinking on the job. His son's toy race cars were strewn across the dash, and his floorboards were covered with a rat's nest of papers and plastic bottles.

Harry saw everything as a sign from God – even us standing on the side of the road. He had nowhere to go and nothing to do, but something spoke to him and said, "Pick them up, and take them where they want to go." Even though he had no reason to drive in our direction, he drove us four hours north.

When his son was a baby, Harry told us he'd watched a news report of a series of crib deaths, and he couldn't sleep for three straight days because he was so worried about his son dying. He'd started to have a nervous breakdown, and he'd known it was only a matter of time before he would descend into inescapable madness.

"I saw smoke billowing in the corner of my son's room," Harry said gravely. "This smoke kept getting bigger and bigger, expanding.

"The smoke took the shape of a giant marble foot, as large as the sky in front of us." He waved his hand at the expanse of the wide gray sky through his windshield.

"And I hear this giggling. It's a baby giggling. And I realize that this foot is the foot of God. And suddenly I see this baby . . ." He paused. I looked over at him. Large tears were rolling down his cheeks. "I see this baby next to this foot. And it's laughing and bouncing up and down. And that's when I knew that all babies go to heaven. And from that day forward, I could sleep."

Before he said good-bye, he told us how he was going to murder his ex-wife and her lover: "I'm going to back my truck into their trailer and finish them off with my shotgun. Then the police will come and we know what will happen from there."

Brent and his wife, Paula – hauling a pair of horses in a trailer behind their battered SUV – would pick us up in Williamston, North Carolina, and take us all the way to Manns Harbor (along the Atlantic coast), where they lived in a trailer on a peninsula near the Outer Banks. They looked leathery, with faces cracked and weathered by forty years of labor, trucking the roads we hitched along. When we got to their place, they invited us in for shrimp and vodka. This was the best part of a hitchhike: sharing comforts, lives, and stories with strangers. Grateful for their hospitality, we offered to help with chores the next day. Sami combed the horses while I heaved a pile of manure out of the horses' shack with a pitchfork.

On our second night with them, Paula confided that she hadn't spoken with her son, who was my age, in years. Upon dropping us off at Jockey's Ridge on the Outer Banks the following day, Paula struggled to hide her tears. Sami and I said good-bye and walked along the Outer Banks, ambled through the Wright Brothers National Memorial, rolled down the sand dunes at Jockey's Ridge – our packs always a safe distance away – and strolled along an endless line of beachfront homes until dusk when we set up our tent behind a dune.

As Sami slept, I listened to the waves curl and crash into the beach. I thought of the last six months in Mississippi and my last week hitching rides with the burned out, beat up, fed up, but always kind and generous sector of society I'd never gotten to know until now. It was strange, I thought, how it was always the poor who picked us up. Our drivers weren't the type who had happy families and middle-class upbringings like Sami and I had. The shiny SUVs or giant, bus-sized RVs would ride on past, but the worse the rattletrap, the more likely it was to pull over for us. Maybe it wasn't strange at all. They lived lives with two feet planted in reality. Perhaps they didn't hesitate to pick us up because they knew what it was like to be cold and hungry and away from home. They dwelled beneath poverty lines and were undereducated, but they were – in the ways that mattered most – far more civilized than the finely bred and carefully raised, for there is no demographic that has a sharper instinct for empathy than the downtrodden.

While I'd been intrigued with the idea of voluntary poverty and was living some semblance of the tramp life, I could see that there was nothing glamorous about the sort of poverty they lived in. There was so much squalor, so much destitution, so much pain. I saw it in Gulfport, in the crew members, in the drivers.

I couldn't wait to get away from the likes of Terry and Rusty and Harry. I couldn't wait to get away from their terrible stories of drugs and divorce, alcohol and addictions. It seemed like everyone we rode with had some tragic past. I thought: *Is this the real America? Have I just been quarantined in my happy little college and suburban bubble my whole life? And is this my real generation: poorly educated, overmedicated, abused, addicted, indebted?*

Yet amid the garbage of Katrina, and when walking through the ruins of the many people I'd met, I saw that flowers still bloomed, lives still went on, the earth rehabilitated, and people reformed. I began to believe that in America, if you give something the right soil, the right nurturing, and, most of all, the room to grow, revival, transformation, revolution – anything is possible.

• • •

The following day was the first of April—the eighth day of our journey. We were headed to Winston-Salem to see Wake Forest, a college that had a liberal studies program I'd read about and thought I might someday apply to.

Sami had become uncharacteristically quiet. I kissed and hugged and joked with her as we waited for rides, trying to snap her out of her sullenness as much for me as for her. It was unsettling to see her so taciturn. Every time I'd ask, "What's wrong?" she'd look off pensively and remark, "Oh, nothing . . ."

We were on Highway 64 holding a sign that read RALEIGH. We got a ride. Then another and another. But we were still far from Wake Forest, so we spent the last moments of daylight setting up camp under a thorny, pink-flowered tree in the woods by a farm that we had snuck onto.

On our backpacking stove, we made one of our standard nightly feasts: macaroni and cheese with ramen noodles. Normally our spirits would be lifted after finding a safe camping spot and eating a warm meal, but Sami was unmoved. We squirmed into our one-person tent, and I lovingly draped an arm around her belly, nestling my nose in the soft curls by her ear.

She took a deep breath and asked plaintively, "So do you want to hear what's been upsetting me?"

There was something mysterious in her voice. Uninterpretable. For the first time, I felt I had no control at all in the relationship. What she was about to say would rattle me to my core.

She began slowly and whispered into my ear: "I haven't been to a doctor but . . . I know. I'm pregnant."

11

SON

Spring 2008–Niagara Falls, New York
DEBT: $11,000

THOUGHTS OF FATHERING A CHILD had yet to enter my head. It was just unthinkable. It was one of those things that I knew would never happen to me, like dying, growing skin tags, or wearing a diaper again.

So, on the eighth night of our hitchhiking adventure, when Sami told me she was pregnant, I was shocked. Not "static on the doorknob" shocked, but "dark angels are swooping down from the sky and the world is coming to an end" shocked. I had so many dreams: dreams to go back to school, dreams to travel the world, dreams to become a mountain recluse. I didn't know exactly what I wanted to do or be, but I was sure I didn't want to be a father. Not yet at least.

Things had been going so well. Because of my frugal, buy-nothing, save-everything lifestyle, I was – despite the low wages – paying off the debt faster than I ever would have imagined. I'd found some balance between work and adventure. Every day I was learning something new about myself and my country, about frugality, poverty, and wealth. And I knew that if I got the right job, by summer's end I might finally be free and out of debt. Really free this time.

But just like that, with Sami's admission, all my dreams were crushed under the heavy belly of pregnancy. I was going to be a father.

In just a few years, I'd be walking around a two-story suburban house wearing saggy, faded briefs. I'd have to work long hours to pay for Xbox video games and Spider-Man Halloween costumes so my progeny wouldn't be ostracized at school. I'd spend my Saturday afternoons – exhausted from my week's toil – in a comatose state, slouched in front of the tube, balancing a can of Michelob on my gut, oblivious to the diapered litter of squealing Ken Juniors bouncing all around me.

Perhaps I had a slightly skewed vision of family life, and I suppose, deep down, I kind of wanted some of the trappings of the American dream: a wife, kids, a home, a car, a basketball hoop in the driveway. But not now. Not when I was so close to paying off my debt.

So when Sami told me she was pregnant, I'd never been so startled, so devastated in my life. It was as if she had told me that someone close to me had died or that I had a terminal disease. My breathing became loud and deep and heavy. I lost control of my exhalations. I started to feel dizzy. The walls of my tent blurred. I thought I was going to pass out.

Sami, concerned, put her hand on my chest and said, "Kenny . . . Kenny . . . Calm down. Calm down. It's just a joke. It's April Fools' Day."

April Fools' Day? Is this her idea of a joke?

I couldn't talk or move. She apologized over and over again above my dead, lifeless body, until she broke out into tears. It was the first time I wondered: *What am I doing with this girl?*

Deep down, I knew our relationship was unsustainable. As well as being kind and sweet and loving and free and wise beyond her years in her own weird way, Sami was also part crazy, part oblivious, part absentminded. She would say and do this sort of stuff all the time, sending me into states of anger and jealousy and frustration.

Because she'd spent so much of her youth on prescription drugs and in hospitals, she'd missed out on a lot of the social lessons we normally get from places like school and at home. This was a blessing and a curse: a blessing in that she got through her teens without being homogenized by sprawling social institutions – as places like high school are wont to do – but a curse in that she missed out on the many lessons that would have helped her navigate through the treacherous norms and customs and rules of society. She'd just woken up from a long and disturbing sleep to a bright and cheery, though uncertain and unknown, world. She was an alien dropped off on planet Earth with no real notions of modern norms and etiquette and convention.

She was still recovering from bulimia, suicide attempts, and the trauma of depression. Whenever she got a stomachache, she needed to be spoken to in soothing tones so she didn't have a panic attack and have to be taken to a hospital. She accepted her problems as if they were hardships she'd have to carry for the rest of her life, oblivious to the fact that self-reformation could be achieved through the process of becoming self-aware. Suddenly, I'd found myself in a situation where I was no longer focused on my development but on someone else's. I took it upon myself to get her up to New York alive, but also to help equip her with the tools so she could begin to self-repair.

For better or worse, she daily reminded me of my lust and love and jealousy and bliss and anger. She reminded me that I wasn't simply a goal-obsessed, debt-crazed workaholic, but that I was warm-blooded and alive. And I loved her with an agonizing fierceness because of it. I was as drawn to her as I was repulsed. Sami, to me, was a wounded bird. And while I partly resented the new responsibility of caring for her, I couldn't help but embrace the sensations of actually having a meaningful role to play in someone else's life. I knew that despite our glaring incompatibility, we couldn't part ways until she was ready for flight.

The next day, I did my best to forget about her joke, and we continued on to New York. We thumbed our way through Virginia, D.C., and Baltimore. We took a train from New York City to New Haven, Connecticut. We ate cider donuts in Vermont, rode a ferry across Lake Champlain, snuck into the hockey rink in Lake Placid where the 1980 U.S. Olympic hockey team beat the Soviets, and watched an Amishman plow his field behind a team of horses in central New York. After twenty-five days of exploring the East Coast, we made it home.

When I was just a couple of hours away from my parents' house in Niagara Falls, I called them up to confess that I'd been hitchhiking and that I was bringing my girlfriend home with me: a pairing of news that would put my poor mother in a state of shock similar to what I'd just felt.

"WHEN ARE YOU GOING TO GROW UP AND START ACTING LIKE A HUMAN BEING?!" my mother screeched, causing me to pull my head away from the phone like a pitcher dodging a line drive.

By now, the poor woman had had it with me going to Alaska and ghettos, on hitchhikes and voyages. She wanted an ordinary life for her son, for her benefit and mine.

Every once in a while over the past couple of years, when we'd talk on the phone, she'd remind me that the border patrol for the U.S.-Canadian border was still hiring and that I could easily get a job back home if I wanted. My dad, too, would call me up and ask, in a half-serious, half-facetious tone, "So when are you going to get a real job?" Sometime in between "real" and "job," I'd picture myself as Charlie Chaplin helplessly transported from one end of a factory to the other over a series of conveyor belts and gears. I used to take the question seriously, but after traveling the country and being with Sami, I no longer thought about careers, or starting a family, or buying a home.

And now that I was back home, I was reminded of the consequences of my parents' relentless toil, which certainly didn't change my impression of "real jobs" or the "real world." My dad, a factory worker, had little to no feeling in his left hand, which resulted from carpal tunnel syndrome from handling factory machinery all his life and from a spinal injury he sustained when he got hit by a drunk driver on his commute to work. He wore a winter glove when driving because the car's vibrations caused him excruciating pain. My mom limped from an ankle injury caused by thirty-five years of standing as a nurse. When she walked up the stairs, she gripped both handrails, gliding her hands along them before picking up her leg and placing it on the next step. I looked on in agony.

My mom would complain about her ankle, and when I'd ask her why she didn't quit her job or reduce her hours, she'd tell me about all the bills that needed paying and how she couldn't afford to lose her health insurance.

God, I was so happy to have them as parents. My father showed up at every hockey game I ever played. My mom was constantly nervous and worried and meddlesome, but only because she cared.

I couldn't take this lifestyle anymore. I was disgusted with the demands of this "real world." *This isn't for me,* I thought, looking around me. I couldn't live the lifestyle they want me to live or work at the sort of job they hoped I'd work at. It wasn't that I had anything against work. After my time in Mississippi, I knew that I could love work. I loved it even though the pay was poor. And I loved it despite the sore backs, the long hours, and the responsibilities of looking over a crew. It was the first time I did something undeniably good: I helped young people and cleaned up the environment. I learned that when work is meaningful and when the worker provides some useful service or produces some useful product, work is no longer "work" but an enriching component of one's day. I had no problem with the idea of a sixteen-hour workday; it was just that I couldn't stomach the idea of someone else deciding everything from my salary to my time off to whether I had health insurance to when I could retire. I wanted to work. But I wanted to be free, too.

On my second day back home, my mom sat down next to me when I was watching TV in the family room. She'd been upset ever since I got home, and I could tell by her rigid demeanor that she'd planned out the forthcoming conversation, perhaps hoping to address serious issues with subtlety and grace. But now that it was time to have the conversation, the deluge of emotions made her leap straight to the point, blurting out, "Ken, is there something wrong?"

"What do you mean?" I asked, taken aback.

"Do you want to die, Ken? Do you think about killing yourself?"

"What? No! Of course not," I said, confused. "What are you talking about? That's ridiculous."

"Then why do you keep doing what you're doing?! You're going to get yourself killed, you know."

"I'm sorry that I hitchhiked again, Mom. And I'm sorry you're upset. But it's safer than you think."

"Ken, promise me you'll never hitchhike again."

This was less a request than a demand. I leaned back on the sofa, my eyes still on the TV. I wanted to submit and say, "Okay, Mom, I never will," but I couldn't. It seemed that for my whole life I'd been feeling guilt for doing – or for wanting to do – what my instincts begged that I do.

How could she think I was suicidal? I'd never been happier. I could say for the first time that I loved my life.

"Mom . . . ," I said, "I don't have plans to hitchhike again, but I can't promise you that I'll never do it again. I plan on living this way for a while. I'm sorry if you disagree with the way I'm living."

"Ken, I think you may just be depressed – "

"No, Mom," I interrupted, "I'm not. If anyone's unhappy it's you. I love you and Dad, but I strongly disagree with how you guys live."

She asked what I meant, but I was flustered, so I could only say, "It's all about work."

"That's true," she conceded. "But, Ken," she continued, pausing to dab her eyes, "I think I'm going to have to start distancing myself from you before something happens."

I remained on the couch after she got up and left. There was no conciliatory hug. No words of comfort. No apologies. I was unsettled knowing that my actions would hurt other people, but I no longer wished to have a diluted life, made faint by living according to the norms and values of an older generation who'd forgotten what it felt like to have the impassioned representatives of soul and spirit lobby their vessel with an unrelenting persistence to take them on an adventure.

Sami took her flight to Alaska so she could start her job cleaning rooms at Coldfoot Camp. I was momentarily unemployed, wallowing in that ghastly stretch of time in between jobs when savings diminish and your debt silently grows with interest like an undiscovered tumor.

Without anything better to do, I thought I'd explore my hometown.

Wheatfield, New York, is a twenty-eight-square-mile rural-turned-suburban community that's home to eighteen thousand residents (95 percent of whom are white), situated on a plot of flatland between the boneyard industrial cities of Niagara Falls and Buffalo, just a couple of miles from where the environmental disaster of Love Canal took place in the mid-twentieth century.

For the most part, the town has kept off the national radar except for a couple of instances, like in the early 1990s when John Wayne Bobbitt, a graduate of my high school, had his penis cut off and thrown into a field by his abused and tortured wife. There was also a brief flare-up in 2007, when PETA got upset with a local who'd mixed antifreeze into a can of tuna to kill a skunk but ended up killing two of his neighbors' dogs, and, later, a bit of a hubbub when a zany developer proposed a

$788 million theme park called the "Magical Lands of Oz," complete with a Munchkinland Waterworks, Uncle Henry's Barnyard Petting Zoo, and the Labyrinth of the Nome King, which – if it hadn't been struck down by my townsfolk – would have been built a few blocks from my home. (Thank you, townsfolk.)

When my family moved to Wheatfield in 1989, I remember a mostly rural landscape: vast fields of waving weeds, rows of papery corn, spring-green adolescent forests, and long, straight-as-a-ruler country roads. In the undeveloped lot next to my family's home was a pond where my brother and I would ice-skate in the winter and catch frogs in the summer. We built forts in the woods behind our house and played hockey in the streets.

Over the past couple of decades, though, the town and our neighborhood have changed considerably. The fields have been smothered with asphalt, and the forests have been yanked out to make way for new subdivisions. Between 1990 and 2000, 1,318 housing units were added, and since 2000, the town's population has increased by 21 percent (or three thousand people). It is one of the fastest-growing towns in all of New York State.

Yet there's nothing unusual about Wheatfield. Across the United States, between 1982 and 2001, 34 million acres of forests and farms, wilderness and rangeland (the size of Illinois), have been disfigured into "developed" land. Wheatfield, I guess you could say, is capitalism run amok, a libertarian utopia where the golden gods of the Free Market and Private Property have reigned mostly unchecked for years. My subdivision, once a cozy hamlet surrounded by cornfields, is now just one small cell of an uncontrollable cancer. The place began to have a dark, creepy, uninviting feel – sort of how I imagined Disney World after-hours: it might be fun for a weekend, but anything more would be like being stuck in a nightmare you can't wake up out of.

Most of the houses had been constructed within the last twenty years. They looked fresh and trim and sturdy, placed squarely on simple, smoothly-shaven monocultural lawns.

Almost all the subdivisions I saw were named to evoke some image of pristine natural beauty that existed there before the suburb: Woodstream Landing, Country Meadows, Woodland Estates, Stone Ridge, Wildwing Preserve.

From my home, I could see suburbs in all directions. When I was a boy, at least there were pockets of woods to stoke my imagination. But now all I could see were endless rows of cookie-cutter homes, bland corporate parks, vast retirement complexes, all separated by a grid of loud, fast, angry roads. The suburban landscape, before, had never produced any thoughts in me or incited any ire, but now, having roamed the Brooks Range, the Canadian wilderness, and the Mississippi jungle, I could imagine the terrible genocide of trees and swamps and fields that took place here years before. We got rid of all that? For *this?*

Sometimes, we can't help but assume the nature of the landscape we inhabit. Just as the farm fosters industry; the desert, frugality; the mountains, hardiness; and a rocky coastline, a romantic restlessness; so does the suburb foster boredom, conventionality, and conformity. I was beginning to assume the shape of the land around me. I'd begun to revert back to high school Ken. I was bored and purposeless. I'd literally gone days without having left the house. I wallowed in self-pity, ate five meals a day, watched marathons of *Seinfeld* reruns, and indulged in an endless enjoyment of naps, masturbation, and sleeping in until two in the afternoon. My stomach got so soft and doughy I thought I could twist my fat into balloon animals.

I spent hours playing video games. Not having played a video game in years, I was impressed by how much the graphics had improved. In these Tolkienesque worlds, I could make out the veins in the characters' muscles; there were elaborate constellations of stars, intricate details of individual plants, and mountains in the distance that weren't just paper backdrops but geographical features you could explore with your character.

It's interesting how many games today take place in preindustrialized medieval worlds (*Skyrim, World of Warcraft, The*

Legend of Zelda). It seems these worlds in video games have become our new frontier. These are the places where we go for adventure. They are refuges of virtual wildernesses – protected plots of pixelated land that must exist in fake worlds because we've denuded and defanged so much of the wilderness in our real world. No one thinks of wilderness in western New York. Because we have nothing that bears the slightest resemblance to wilderness, we are as unaware of its existence as we are unaware of some undiscovered alien race.

The video games offered mild amusement, as they always had done, but I knew there was something ridiculous about exploring a fake world on the television screen while sitting on a couch in the real one. I needed to get out of my funk. I needed to get out of the house and go to someplace wild. I thought it might do me some good to again view the storied cataracts I'd visited countless times as a boy but never really appreciated.

While Niagara Falls the city may conjure nostalgic images of the idyllic honeymoon capital of the world, today the town is mostly boarded up and deserted – deserted except for a silvery sky-high Native American casino that's positioned in the middle of town, which delivered a fatal tomahawk chop to the neck of most of the small businesses below.

I looked at the falls and felt nothing. There was an ocean of water hurtling over the cliff, dashing onto the rocks below, creating a cool, hazy vapor that left beads of dew on the tourists' hair. The scene was impressive, but only in the way a building or a city is impressive. I couldn't see it, but I could sense that this place had been manhandled. Niagara Falls was damaged goods. In decades previous, the river had been dammed so engineers could alter the rate of erosion. Bolts were put in place to strengthen faults. Retaining walls had been built that removed four hundred feet of the falls. At night, floodlights shone on the falls to turn it different colors, as if the terrifying force of 6 million cubic feet of water per minute needed augmentation to earn people's appreciation. It was surrounded by wax museums and chocolate shops and the towering casino, not to mention the curling tendrils of civilization: guard rails, phone lines, street cables – a tangle of technological intrusions that disallowed its spectators from feeling the sublime, the holy, the transcendental.

I watched how the water – a mass of individual droplets, each once wild and unruly and free – was being contained and controlled, channeled into hydroelectric power, gawked at as a gaudy spectacle. This place had been uglified, commoditized, urbanized, civilized. This place, I decided, was my old self. Here, like Niagara Falls, I'd been deformed and disfigured by forces outside of my control. I'd been bent into a consumer by TV, molded into a conformist by schools, and made into a loan drone by a hundred other things. I'd been paved over, polluted, and planned out. I'd been Love Canal-ed. I'd been civilized.

I never understood why I'd wanted to drive to Alaska so badly years before. But I did now. I realized that I'd needed to see some place that was real, some last corner of the world that hadn't been buried beneath strata of pavement and people, technology and trash. Of course I thought civilization, with its vaccines and bug sprays and its libraries and theaters, was real and had value, too. But part of me knew I needed to escape it for just a bit, to see someplace wild. And maybe, hopefully, if I lost myself in a place like the Alaskan wild – a place where the land might be what it's supposed to be – I might begin to see how I'm supposed to be, too.

Now that I'd been there and back, Niagara Falls, this place, this home, no longer felt like home.

When I got a call from the Gates of the Arctic National Park telling me that I got the job, I knew I was really going home.

12

········

RANGER

Summer 2008—Gates of the Arctic National
Park and Preserve, Alaska

DEBT: $11,000

W**ILDERNESS IS A WELLSPRING** of wild dreams. Leave the forest on the edge of your suburb unexplored and the place will expand in your imagination. It will assume a mysterious, enchanted nature, growing into something like a wilderness. Even though that forest may actually be half-diseased and carpeted with an understory of used condoms and crumpled beer cans, for all we know it could be a howlingly wild, green-bearded Germanic forest where tree nymphs gather in glades and mythic beasts live out great stories. Perhaps it's best to leave those few remnants of nature unexplored. Just as it's good to have nearby fields to supply us with food and aquifers to catch and clean our water, it's good to have a place nearby to supply us with dreams. If we set foot inside our only meager forest and walk its perimeter in an hour, we'll see it for what it really is, and the wild dreams it once created will never be dreamed again.

The Brooks Range, though, can neither be known in a day nor a thousand lifetimes. It is a land of such wildness and breadth that when you're hiking through it, its absolute unknowability will impress itself on you with a force that can steal the air from your lungs. It is a borderless, boundaryless country, an expanding universe whose farthest reaches are always beyond the grasp of your imagination. When hiking through its river valleys or over its mountain passes, you begin to feel that Earth is not limited by spherical contours, but that this planet is the flat-bottomed bedrock of the universe around which all else revolves. Do not come to the Brooks if you wish to receive an education in the finiteness of space. Come to be staggered by a sense of infinity. The Alaskan arctic is impossible to fully explore, to fully know. There's always another creek to follow, another promontory to gaze from, another dripping-wet cave to climb into.

It's a shame that there aren't more places like the Brooks, like a vast midwestern prairie where bison can roam, or a southeastern forest of pine and maple set aside for a thousand years, so, at the very least, the feelings that a vibrant food chain and an ancient forest can conjure could again be enjoyed by man.

Let the intrepid few have wild places where they might be stunned by the extraordinary, and let the sedentary many have such places, too. Though inaccessible for them and the cumbersome, vehicular shells they carry, the forests, mountains, jungles, and prairies will assume grander forms if we give the romantic hand of imagination a chance to color in the blank spots of the map — as it is wont to do — with deeper caves, larger peaks, and secret mountain lakes. Perhaps we'll imagine such places inhabited with bears and lions, sea monsters and Sasquatches — all of which may not exist in real life, but they, if given a place to dwell, will visit and enliven our dreams.

The quicker we Google Map the earth, the more we restrict our native planet from evoking feelings of wonder and enchantment and love for it within the hearts of its human inhabitants. Earth should always remain partly unknown, partly undiscovered, partly unclassified. It will be a tragedy if the Brooks Range comes to be covered in tar and tourists like most any other national park. A vast chain of mountain peaks that can only be traveled by foot radiates a rare beauty, one that is there for everyone to feel even when it cannot be seen.

Many do not know that the songbirds we hear singing from branches on the family aspen in Wisconsin or the street-side oak in California are natives of the north, having been born in the exuberance of an arctic summer. And just as we may forget how the arctic is a birthplace of birdsong, we may fail to appreciate how many of our dreams have been born in the north.

In the Brooks – in each distant copse of trees, in each faraway pond, on each unscalable peak that we'll never see or smell or climb atop of – is someone's dream, for we are enchanted most by the places we can never go. While the Brooks may never provide us with resources for energy or grounds for settlement, so long as they remain wild, they'll forever serve humanity as a factory of dreams.

I was glaring out the window of a three-seated Cessna 185, flying above the Gates of the Arctic National Park. Kurt, the pilot, tilted the yoke and tightly circled a pair of lounging gray wolves that would have been running for their lives before this was a national park.

Kurt dropped the plane just feet above the rocky gravel bar next to the North Fork of the Koyukuk. At the last second – just before it seemed we were going to miss the bar and belly flop into the icy, clear river – he flicked up his wings, increased speed, and tilted the plane up into the air to seek a better landing site. My stomach's contents tumbled like clothes in a dryer. I squirmed to retrieve the plastic bag from my back pocket, nervously chanting, "Happy, happy, happy," wishful for ground under my feet and an outhouse I could high-step to. This was how I got to my job every week.

I was one of five backcountry rangers at the Gates of the Arctic National Park, the northernmost and, at 8.5 million acres,

second-largest national park in the country, as well as, argu-
ably, the wildest left in the park system. About the size of Mary-
land, the "Gates" is situated on the same latitude as Coldfoot,
just above the Arctic Circle. (Add in the abutting Kobuk Valley
National Park and Noatak National Preserve, and the acreage of
the contiguous parkland is slightly larger than West Virginia.)
Because there are no roads, trails, or facilities within the park,
we had to be flown in on bush and float planes.

Before the season started, we had two weeks of training: bear
encounter training, bear spray training, shotgun training, canoe
maneuvers training, and dunker training – the last of which
required that we escape from a simulated cockpit overturned in
a swimming pool.

Our job duties were simple. We'd go on eight-day wilderness
patrols with a fellow ranger, either by foot or canoe. Basically,
we were to follow a route determined by our supervisor, talk
to visitors about leave-no-trace ethics and educate them about
bear safety, including ensuring they stored their food in bear-
proof "bear barrels" (though it would be rare to come across
more than one group on a given patrol), and clean up any trash
or "human impact" that we might find. It was our job to be the
eyes and ears for the law enforcement rangers whom we'd call
on a satellite phone if we ever happened upon poachers or people fishing illegally.

To some, backcountry ranger may sound like a dream profession (and in most cases it was), but it bears mentioning that
we did have to work in conditions that other Americans would
deem inhumane at worst and masochistic at best. When shoul-
dering a sixty-pound pack, being acupunctured by the needles
of a thousand mosquitoes, stumbling over never-ending fields
of tussocks, and always worrying about the possibility of getting
run over by a moose or mauled by a bear, you sometimes think
that such conditions might get you sympathy from Sherpas,
pyramid slave laborers, and third-world textile workers. But,
for the most part, it was a job we were all grateful to have.

We were all male, in our mid-twenties (except for one middle-aged ranger), and paying off our student debts. Each of us had his own method: While I was determined to get rid of mine quickly, Adam, twenty-eight, decided to make the lowest payment possible and stretch his debt out over decades so he was encumbered with only a small monthly payment, allowing him to be, as he put it, "pre tired." Dick, twenty-three, had just graduated from the University of Alaska at Fairbanks with $70,000 in debt and decided that he'd take out another loan to buy a plane so he could land a better-paying job with the Park Service. Tomas, the volunteer, was still accumulating debt.

At the beginning of the season, we rangers excitedly talked about the places we might get to see on patrol. Perhaps the ostentatious Mount Doonerak, a 7,457-foot spire thrusting itself up and over the nearby hills like a solitary hand raised in a classroom. Or the Arrigetch Peaks, a ring of jagged, thorny, pointy, sheer-walled granite cliffs, whose bony fingers and vampire nails reach upward, clawing the sky. The Valley of the Precipices. Oolah Pass. Rivers and lakes pronounced with a click of the tongue and hushed respect: Itkillik. Takahula. Itikmalak. Tinayguk. Alatna. Noatak. Agiak. Unakserak. Kurupa. Nigu.

After I got dropped off on the edge of the North Fork of the Koyukuk River for my first patrol, my first thought was: *This is too good to be true.* My second: *Why did they hire* me? I didn't want to already question the integrity of my new employer, but I mean, hell, just three years before I did sorta get lost in this very park.

I looked at the river. The Koyukuk writhed down the valley, gushing between two of the Gates's grandest peaks: Frigid Crags and Boreal Mountain — the very "gates" to the arctic that explorer Bob Marshall named in the 1930s. My ranger-friend Ted and I were to float down the Koyukuk and explore the surrounding country as we pleased. We overturned our inflatable canoe on the riverbank, tied it up to some heavy logs, and set up camp next to the river. We filled our water bottles with river water and set out to climb a mountain from which we'd hoped to spot a poacher or a group of hikers or something that would make us feel purposeful. But all we'd see was a never-ending chain of mountains of a Jurassic landscape that bore a beauty that repulsed as much as it allured. The distance from the security of civilization evoked in us a sense of urgency for the comforts and safety of home, yet the terrifying grandeur of an endless wilderness invited us to embrace the wild and unruly sensations of the uncertain now.

While on the mountain, a freak gale swept down the valley that caused us to teeter on the ridge and – unbeknownst to me – my tent, below, to fly off the ground and into the air like a beach ball. Hours later, when Ted and I descended the mountain and got back to our camp, I frantically scanned the terrain for my missing tent. I hardly even thought about the very scary possibility of being helpless and without shelter in the middle of the wilderness. All I could think about was that I'd forever be remembered as the "guy who lost his tent on his first patrol" and having my lost item report greeted with raucous laughter when it was shown on slideshow presentations at staff Christmas parties. I walked down the riverbank hopelessly. *Please, please, please, let me find my tent!* And there it was, half-submerged in an eddy downriver. Thankfully, it hadn't gone any farther because my copy of Ayn Rand's corpulent opus *The Fountainhead* got soaked, sinking the tent to the bottom of the river floor as if it were a treasure chest in the hull of a mighty ship. (Perhaps the only good the book has ever done.)

That night, I stepped out of the soggy tent to use the bathroom. I was squatting in between two spindly rough-barked spruce trees. That evening, everything – the trees, the mountains, the air – was coated in an august gold. The hue of the atmosphere portended a vicious storm that would strike in a matter of minutes, but for that moment it was nothing but serene beauty; it was as if the sun had exploded into a trillion radiant particles. A large bull caribou loped out of the bush and trotted past me, unaware of my presence. There was a moment like this on every patrol, when I, overwhelmed, could only shake my head and think: *What an incredible world.*

On subsequent patrols I'd saunter around the narrow-as-a-rifle-barrel Walker Lake, whose only island was populated by an abandoned caribou calf and a frenzy of swallows that were building mud nests under the roof of a deserted cabin. I'd float through the narrow red canyons of the deliciously clear Kobuk River, populated with grazing grayling and prowling sheefish. I hiked seven days north of the tree line from the Inuit village of Anaktuvuk Pass to the Dalton Highway through a corridor of treeless, pyramid-shaped mountains.

Every day there was a new animal sighting: inquiring beavers, waddling porcupines, stoic wolves, a tribe of Dall sheep that looked like specks of snow on green mountain peaks, a sow and two grizzly cubs sprinting away upon catching our scent, intrepid harlequin ducks, monstrous trumpeter swans, ghostly snowy owls.

After my fifth patrol, I'd begun to figure out how to travel in the arctic. I'd made enough mistakes to be leery of again falling victim to the vices of hubris and machismo. I knew that if I wanted to make it out of the arctic alive, then I'd have to remain smart and humble and flexible, always open to learn new lessons that the land could teach me. I could now "read" the mountains better, so I knew where I could find the hardest ground and the most tussock-free path. When I crossed rivers, I knew to face into the water's surge. I stored food in bear barrels downwind so bears wouldn't be led into camp, and I became familiar with the tracks of all the large mammals, as well as many species of birds and varieties of plants.

Not only was I becoming intimately acquainted with the Brooks, but I was getting paid to do it. And my god was I getting paid. I'd never seen so much money. I was getting paid $16 an hour, plus a 25 percent cost-of-living adjustment (which we received because we lived in a remote, expensive area), boosting my pay to $20 an hour. This was an unbelievable amount of money. It was an *absurd* amount of money. Part of me wanted to give half of it back. Before this, I'd never gotten paid more than $9 an hour.

I had more debts to pay, though. I'd borrowed $5,000 from my mother so I could buy a flight up to Alaska and a summer's worth of food, which had to be bought in Fairbanks and driven up to Coldfoot. I also had to get a car so I could travel between my new home (situated five miles north of Coldfoot) to the ranger station in town. I spent about $600 for the flight, $600 on food, and $3,000 for a 1999 Dodge Stratus. Working at the Gates was a large financial investment, but I knew it would be worth it by summer's end.

I'd still be making my standard $114 monthly payments on my government student loan, but my main priority was to pay my mother back, if just to simplify things for my psychological well-being. (Better to have one debt than two. Simplify, simplify, simplify.) So every two weeks I sent my checks home, always keeping close tabs on my online bank account to be sure my money was going where it was supposed to go.

By the end of June, I'd paid my mother back. The car, the food, and the flight were mine, paid by me. And then I put my full effort into what was left of my undergraduate debt. Every two weeks, I'd missile another check at it, gleefully watching the debt disintegrate.

$11,000. $9,500. $8,000. $6,500. $5,000. $3,500 . . .

Josh, meanwhile, who was still more than $50,000 in debt, continued to do "admissions advising" for Westwood College. On a normal day, he'd make 150 calls to prospective students who had made the grave mistake of typing their phone number into some online questionnaire, hoping to receive information about college options. Within days, that phone number would be sent out to a handful of for-profit colleges whose admissions representatives — like Josh — would call every day for the next couple of weeks.

On his commute to work, Josh would often be brought to a halt by the rush hour traffic. He'd glance at the Rocky Mountains to the west, remembering how he'd just been in the Brooks Range climbing mountains and living some semblance of a free life a year before. Once inside his building, he sat on a swiveled chair in a cubicle, wearing a headset in front of a computer, which was faced toward the aisle so his superiors could make sure he was working.

Westwood reps were taught about a variety of the school's programs, the career center, and the supposed value of the education, but a few crucial details were left out. The reps, for instance, knew little about the graduation rate or about how many graduates were getting jobs in their academic fields. The reps were, however, given a thorough education in sales. They were trained in the "the seven-step sales process," the "cookie close," the "sandwich technique," and the "takeaway": emotional mind games, as Josh described them, that he was supposed to use on the students he'd talk to. It was his job to get as many kids to sign up as possible.

The most successful admissions reps would get huge bonuses, paid time off, and even trips to Cancun. It began to dawn on Josh that something wasn't quite right with the place when, at an office party, a Best Liar award was given by an assistant director to a rep who was notorious for using the most ruthless methods to persuade kids to enroll. The office smiled and laughed and clapped for the winner.

He tried to do his job as ethically as possible, but because he wasn't signing up enough students, he'd been put on probation, kicked off his sales team, and warned that he might lose his job. Because of his debt obligations, he had to begin to think about himself. Now, when he pitched the school to students, he avoided telling certain details. He began pushing the right emotional "pressure points." His supervisor, impressed, told Josh that he was beginning to meet his potential.

Josh could have given Westwood a black eye if he went public with what he knew, but he felt powerless. He needed the job.

He couldn't go back to being unemployed. This was, after all, the first time he was paying off his debt at a reasonable pace. Josh thought about quitting and issuing some sort of Jerry Maguire–like mission statement to all his coworkers to redeem his soul and undermine Westwood, but the debt was getting the best of him.

I could see that Josh was "becoming" his job. The job was making him into a tool, a machine, a piece of equipment. He convinced young kids to go to school so that they could be tools and equipment and machines, too. He was a debtor, in the business of putting other students into debt. The job spread over him, like a suburb over wild land.

My job was seasonal. It started in May and ended in October. I spent half of my season in the backcountry and the other half in Coldfoot at the ranger station, where I worked behind the counter greeting visitors, performing backcountry orientations to would-be hikers, and occasionally giving a slideshow presentation on "Literature of the Arctic."

When I arrived in Coldfoot after a patrol, I'd get out of uniform, shave off a week's worth of stubble, and scrub myself clean in the shower. Sami, who worked as a lodge cleaner at Coldfoot Camp, would come over to my house, which I was renting from the Park Service, where there'd be passionate embraces, kisses, sincere I-missed-yous.

I had an adventurous, well-paying job. A loving girlfriend. And my debt was almost paid off. Years before, I would have thought that these would be the very ingredients for a happy, fulfilled life. Perhaps they'd make me want to settle down. Yet I was troubled now that I had them. I was beginning to feel at ease. Too at ease. The walls were closing in around me. I was being crowded in by sensual pleasures and soft comforts: a warm home (when I wasn't on a patrol), a steamy romance, a stable job, an adequate salary. My supervisor hinted that I had long-term potential and that I should go to law enforcement school to get certification so I could land a permanent job with the Park Service. Part of me was enticed by the idea of having a noble job and a steady paycheck, and part of me felt lured by the sultry Jacuzzi of drowsy-eyed, vapid material comfort that I'd be able to indulge in. But I knew I couldn't settle into a career — not yet at least. I knew I couldn't allow myself to feast from the buffet of domestic pleasures, either. As much as I was a disciplined voyageur, an intrepid hitchhiker, and a stalwart ranger, dwelling in me also was an unmotivated suburbanite, a lazy couch potato, a pitiful, sometimes alcoholic, loser. If I put myself in a comfortable situation, my lesser self would take over, emerging from his dark, refurbished basement cocoon with disheveled hair, wearing a tattered, loose-fitting pair of sweatpants, and announcing that he'd be instituting a new policy of unambition and sloth.

I realized the journey that I was on wasn't about getting out of debt or finding my perfect job, or girlfriend, or life. It was about becoming the best man I could be. I knew from experience that the only thing keeping me from turning back into the person I'd left behind in New York was voluntary simplicity.

And I knew I needed to undergo another period of rigorous training. As inspired as I was in the Brooks, and as honorable as I thought the ranger profession to be, I was little more than a paid hiker. I thought of a Saint Francis of Assisi quote. He said, "He who works with his hands is a laborer. He who works with his hands and his head is a craftsman. He who works with his hands and his head and his heart is an artist." If I was a laborer at the Home Depot, I was a craftsman at the Gates. Now, it was time to learn how to be an artist — someone who used more than his hands and head for work.

When I was on patrol, most of the time a fellow ranger and I would paddle our canoe in silence. I had almost the whole day to let my thoughts meander with us down the river, though they always got caught up into two swirling eddies: 1) I never, ever want to go into debt again; and 2) I want to go to grad school. I want to be a part of a university.

. . .

At the end of the summer, on my last day at Coldfoot, in the early hours of the morning, I said good-bye to Sami, who was now working as a waitress in the café. She kept delaying my departure by bringing me food and watching me eat it.

"Sami, I have to go," I said. We both knew I was more than just leaving. This was good-bye.

Tears welled and her face crinkled.

"I don't want you to go!" she exclaimed, clasping my torso.

I didn't want to go, either. Oh, how I wanted to stay and nestle my head in betwixt the tender breasts of love and lust, and keep it there, and stop worrying about goals and journeys and self-improvement. But I was headed to Denver, and she, to California, where she'd enrolled in a community college. It was the season for selfishness, or at least that's how I justified it.

What had our relationship been? It was about to fall to the ground in a rush of dust like a demolished building. It was a young cottonwood about to be charred and splintered by an errant lightning bolt. It was a wild river doomed to be dammed. But I didn't feel like anything or anybody was being ruined. Our relationship was kind of like our hitchhike adventure. It wasn't supposed to last forever. It was simply a means of getting from one point to another, of helping us get from our past selves to our future selves. And like the hitchhike, we'd take from it, learn from it, say good-bye to it, be the better for it, and think back on it fondly.

Devastated, and completely unsure about what I was doing, I kissed her forehead and left the café.

13

....

PACKAGE HANDLER

Fall 2008—Denver, Colorado

DEBT: $3,500

I HEARD JOSH WALK INTO the house. After nearly twelve hours of dressing, commuting to work, working, and commuting back home, his workday was finally over. I was downstairs in his basement, busy comparing the cost of one grad school's liberal studies program to another. Upstairs, I heard the paws of his greyhound, Lois, click along the wood floor to greet him. Josh took slow, halting steps down the basement stairs and dragged his feet to the couch. His dress shirt was only half tucked into his pants and his tie hung around his neck like a noose. He wore black-rimmed glasses, and his hair still stood hard and crusty from his morning dressing of hair gel. He collapsed onto the sofa like a ball into a catcher's mitt, unleashing a plaintive sigh before flicking on the TV and typing his favorite euchre website into his laptop.

"Hey, man," I said.

"Hey."

"How was work?" I asked.

His attention had already been pulled in too many directions. We had once debated great things over e-mail, but now our conversations were rarely any more substantial than this. It was as if Josh had been pulled down the polished halls of industry, strapped to a bed in some padded, soundproof room, and had his spirit exorcized out of him by Corporate America's Nurse Ratched. He was a dull, tired, sleepy version of the Josh I once knew. I half-wanted to smother him with a pillow to put him out of his misery.

He was once so passionate, so idealistic, *so borderline violent.* (When John Kerry lost to Bush in 2004, I remember Josh driving his fist through his bedroom wall.) Now, the only time he ever showed any emotion was during the inevitable Buffalo Bills loss or when he was engaged in his weekly dispute with his girlfriend about laundry. He grudgingly spent his weekends doing home maintenance projects: painting the fence, putting drywall in place, manicuring the grass, or remodeling the bathroom. He wanted to do nothing more than sit and rest on his weekends, but he was obligated to maintain the house and attend to an endless series of his girlfriend's friends' weddings and baby showers. And he just went along with it all. He couldn't not go along with it. He needed his job. He needed his girlfriend. He needed his girlfriend's home. He couldn't simply get up and skip town like I could.

Weeks before, I'd sold my car in Alaska, road-tripped with a fellow ranger down to Denver, and got a part-time gig handling packages for UPS for $13 an hour. I worked during the busy holiday season alongside a black, homophobic driver named Dwayne, who delivered packages in a trendy, largely gay part of Denver.

With the goal of enrolling in grad school, I knew I had to finish off my debt and line my nest with whatever money I could make. So I worked and saved and dreamed.

I tried to reinvigorate Josh with grand ideas. Earlier, he'd told me he thought about going public with what he believed were evils committed at Westwood. But when I reminded him about his idea, he said, "I don't know about that anymore," without taking his eyes off the computer. "I really don't want to burn any bridges."

I had sympathy for him. Without his job, his debt would spiral out of control. He wouldn't be able to meet his payments, and the interest would accrue. At the same time, I couldn't help but wonder: *Who is this person?* This wasn't the Josh I'd known – not my friend who laughed while vomiting at the Yukon River Camp, not the hiking partner who climbed mountains with me bare-chested in the Brooks, not my freshman roommate who wanted to join the Peace Corps and save the world.

What had happened? Did the job do it to him? The debt?

It amazed me how thoughtlessly Josh and I had gone into debt. It amazed me how thoughtlessly we had surrendered our autonomy. It wasn't just us, though. The whole nation was in debt. Going into debt had become as American as the forty-hour workweek, a stampede of Walmart warriors on Black Friday, or the hillocks of cheap plastic under Christmas trees. As a country, we marched from one unpaid-for purchase to the next in a quest for fulfillment that fades long before the bill arrives. I thought about how similar we were to the Spanish explorers who'd dedicate their lives to finding El Dorado – always around the next bend in the river, yet never there at all.

Why do we do it? Why do we go into debt? Why are we willing to take out tens of thousands of dollars in loans?

Looking at Josh on the sofa, I thought that maybe he and I actually *wanted* to go into debt. Maybe a part of us wanted to be constrained and repressed. Maybe we wanted to be stuck in school for four years and, after that, stuck in debt for decades. Maybe we wanted these limitations and walls because they made life simple. We wouldn't have to be bothered by the great existential questions of our day if we had to spend forty-plus hours a week "gettin' 'er done." If we take gravity away from a man, odds are he'll fear the novelty of flight so much that the first thing he'll want is his feet back on the ground. If freedom was our fear, debt was our gravity.

As a country, we take out loans and go to school. We take out loans and buy a car. We take out loans and buy a home. It's not always that we simply "want" these things. Rather, it's often the case that we use our obligations as confirmations that "we're doing something." If we have things to pay for, we need a job. If we have a job, we need a car. If we have such things, we have a life, albeit an ordinary and monotonous life, but a life no less. If we have debt, we have a goal – we have a reason to get out of bed in the morning. Debt narrows our options. It gives us a good reason to stick it out at a job, sink into sofas, and savor the comforts of the status quo. Debt is sought so we have a game to play, a battle to fight, a mythology to live out. It gives us a script to read, rules to abide by, instructions to follow. And when we see someone who doesn't play by our rules – someone who's spurned the comforts of hearth and home – we shift in our chairs and call him or her crazy. We feel a fury for the hobo and the hitchhiker, the hippie and gypsy, the vagrant and nomad – not because we have any reason to believe these people will do us any harm, but because they make us feel uncomfortable. They remind us of the inner longings we've squelched, the hero or heroine we've buried beneath a houseful of junk, the spirit we've exorcised out of ourselves so we could remain with our feet on the ground, stable and secure.

It's especially telling how young people, today, deal with their debts. So many debtors just seem *okay* about it. No one, of course, likes his or her debt, and everyone, of course, wants it gone, but among many debtors there is a curious lack of urgency. They exhibit an almost brazen indifference about owing tens of thousands of dollars, or going into forbearance, or having hundreds of dollars taken from their monthly incomes. Most are fine with going on the twenty-year repayment plan, as if nothing was at stake. They thought of their debt as some annoying, inevitable bill like car insurance, not the steel bars that kept them confined to lives of everlasting obligation.

Every weekday afternoon, in my brown slacks, I'd jog to a street corner where Dwayne would pick me up in his UPS truck. He'd drive us through Denver's business district — largely made up of gay clubs, furniture stores, advertising firms, and mechanic shops. He'd park the truck in front of a building, and I'd scurry into the back of the truck to find the appropriate boxes while he entered in the information on my handheld signature device. Then I'd waddle into the building with a careening stack of boxes as Dwayne pulled up to the next business, into which he'd roll a dolly of packages of his own.

When I got home, I'd tidy up the house, take Lois for a run, or finish painting a fence or some such project, before heading down to the basement to commence my research on schools. I made a list of every graduate liberal studies program in the country. While the quality of the program was of course important, I wouldn't consider enrolling in a program that was out of my price range, which I decided was more than $1,000 a course. I chose to apply to two that fit my criteria: Wesleyan University in Connecticut and Wake Forest in North Carolina. I no longer harbored dreams of attending some Ivy League–caliber school, and I knew from experience that they had no desire to welcome me into their prestigious ranks, but I thought, *What the hell, why not send out one more?* So I applied to the school that had one of the premier graduate liberal studies program in the country, Duke University.

Meanwhile, some late checks arrived from the Park Service, which I put toward what remained of my debt.

$2,000. $500 . . .

And then the day came when I got the check that would finally kill the debt.

$0.

The debt was gone.

Up until the moment I paid it off, I wasn't sure what I was going to feel. I'd been waiting for this day for two and a half years. This day had been in the back of my mind for almost a decade. And here I was, out of debt. Debt-free. . . .

I was, uh, free . . .

I felt slightly apathetic. Perhaps it was because I'd killed the debt so unceremoniously. There was no check I could sign with a triumphant flourish, no repo man I could tell off, no Sallie Mae I could hang up on. There were no fireworks or foaming bottles of champagne – just a pop-up window on my computer saying the debt had been paid off.

I didn't feel like I'd been "freed." Maybe it was because I'd already felt freed. Or maybe it was because, in a strange way, I'd actually felt most alive when I felt most constrained. Something happened that day in my car in the UB parking lot a few years back, when I heard that voice and those three words. Perhaps it had finally dawned on me that I was stuck, restrained, and in debt. Sometimes it's not until you see your shackles that you see your dreams. The soul must first be caged before it can be set free. For all the trouble it had put me through, I had the debt to thank for that.

Still, though, now that I was out of debt, I couldn't stop dreaming about what I could finally do. This sense of hope and anticipation that I felt made living a delight. *This was freedom*, I thought. Freedom didn't have to be about tramping around or having adventures; freedom was simply being able to entertain the prospect of changing your circumstances.

Several more checks came in from the Park Service and UPS. I now had $3,500 in the bank. It was mine. All mine. It was the first time I had had a true surplus of money since I was a thirteen-year-old paperboy.

And now that I was finally debt-free, more than anything in the world I wanted to go to grad school. Yet, having just paid off my debt, the very last thing I wanted was to go back into debt again. And while it seemed like it might make sense to just keep working and saving so I could afford to pay for my education someday, I knew I couldn't do that; if I'd learned anything these past couple of years, it was that a postponed dream was just a dream. If I didn't do it now, I might never.

Yet I wondered if it was possible to pay for a graduate education without taking out loans. Was my original $32,000 student debt inevitable, or could I have avoided it if I'd known then what I knew now?

I saw how much Josh had to pay for the house we were living in: the plumbing, the electricity, the mortgage, the gym membership, the dog food, the NFL Game Rewind subscription. He had to pay for it with his time and his freedom.

Might there be a better way?

In *Walden,* I remembered how Thoreau said something about a six-by-three-foot box by a railroad. At night, laborers would lock their tools in it. A man could live comfortably in one of these boxes, Thoreau declared. Nor would he have to borrow money and surrender freedom to afford a "larger and more luxurious box."

("I am far from jesting," he clarified.)

If I wanted to go to school without incurring debt and losing freedom, I had to think outside the box, too – or, as Thoreau might suggest, inside one.

I wondered: *What if I built and lived in a box of my own?* Perhaps I could slap together a small shack in the woods somewhere close to campus. But then I remembered that I had no carpentry skills whatsoever, and that my ungainly hammer swing in the past had provoked friends to ridicule me with a flagrance that my other deficiencies had failed to arouse.

Okay, no shack.

What about a tent? I could take my one-person tent out into the woods, or, better yet, I could buy a big five-person tent that would give me plenty of space. I fantasized about being a student by day and wild man of the woods by night. I'd let my shoulder hairs grow out, whittle spears, and hunt squirrels barefoot. During full moons, I'd unleash forlorn howls, always draped in a cloak of patched-together rabbit furs. But then I thought: *What if someone raided my tent when I was in class?*

Okay, no tent.

What if I just lurked in libraries, stairwells, and the student union? I'd hide my stuff above ceiling tiles and embrace the role of some wall-dwelling, pale-faced specter, affrighting students in libraries with spine-chilling cackles that I'd transmit through my intimately-known labyrinth of air vents.

But it was more than clear to me that none of these ideas would have worked. I needed something that was safe. Something that would provide security. Something cheap. Some form of housing that wouldn't make campus security suspect me. Something that . . .

I've got it!

James's Chevy Suburban. No, wait, I could do better than that. A van! I'd buy a van. I'd find some cheap piece of junk, buy a parking permit, and convert it into a mini dorm room. I'd take showers at the gym, get electricity in the library, and cook meals on my camping stove. And to ensure I wouldn't be caught and thrown off campus, I wouldn't tell anyone. It would be my secret.

Now, I just needed to find my piece of junk.

I opened Josh's mailbox to see two letters. One was from Wesleyan and the other from Wake Forest. Just as happy events can sometimes make one cry, and traumatic events can make one laugh, I opened the envelopes and looked aghast at the messages. They didn't start off with an "Unfortunately." They didn't say, "We regret to inform you." My god, someone—*two* someones—had actually accepted me.

I was also preparing for a phone interview with Duke, during which the program director and one of the department's founders would ask me about myself and my intellectual interests. I spent two days writing out every conceivable question they could ask and then typed out my responses. I read the responses aloud into a recorder and listened to the recording so I could emphasize the proper words during the actual conversation.

The day of the interview, I drank a cup of coffee, leafed through my papers, and took the call. During the conversation with my two interviewers, I came across as something other than a bumbling idiot. "We'll let you know," they said.

A week later, the letter from Duke came. I took it down to the basement and sat on the couch, imagining what terrible revelations of woe were sealed inside. I opened it up and sighed.

It read:

> Your application for admission to the Master of Arts in Liberal Studies program has been carefully reviewed by the Liberal Studies Admissions Committee and by the Graduate School. I am happy to inform you that you are granted admission to the Graduate School for the 2009 Spring Term.

I leaped in the air and ran over to tackle Lois. I'm going to Duke! I'm going back to school!

Change was in the air. Bush was gone and Obama was in office. Fall was chilling into winter, and I was moving to North Carolina.

In late December, Josh drove me to the airport. We hadn't e-mailed each other in months. And now that we communicated face-to-face, it had been a while since we'd had a real conversation. But I wasn't sure if we'd resume our correspondence. We were heading in such different directions. I'd never felt more idealistic, while Josh had resigned himself to his circumstances. I hugged him and wished him luck on the debt. He handed me a package and said, "This might come in handy."

It was a power inverter that could be plugged into a vehicle's cigarette lighter if I wanted to create electricity in the van.

When I told other people about my plan, some rolled their eyes and others told me I'd get caught. I couldn't even tell my parents for fear of what they might think. But Josh was always supportive of everything I ever did.

Despite the shirt and tie, the loafers, the gelled hair, the boring car, the boring home, the boring life – underneath this landfill of professionalism – I could still see something wild and pure in Josh. There was still some beast in there that would eventually bend the bars that held him inside. I wished I could have taken him with me.

As I got on the plane, I felt poor but free. I had a little over $3,500 in the bank, a backpack of camping gear, and some theories that I needed to test out.

I made it my goal to do whatever it would take to graduate from Duke debt-free.

Part III

....

GRAD STUDENT,

or

My Attempt to Afford Grad School by
Moving into a Creepy Red Van

—Day One of Vandwelling Experiment—

14

PURCHASE

January 2009—Duke University

SAVINGS: $3,517

| FLEW FROM DENVER TO Wheatfield to spend a couple of days with my family during the holidays. My mother, beside herself that her son got into a good school, called up her friends and proudly boasted to neighbors. She bought me a $350 laptop and refused my offers to pay her back. She also was kind enough to provide me with a full inventory of all of Greater Durham's student housing options. I was gracious, but I told her—looking her straight in the eye—that I was "going to buy a van and live in it." That was the sort of news that normally would have evoked a terrified "Oh please, Jesus, no" response from my mother—the sort of response that I surely would have gotten if I'd have, say, started off a sentence with "Me and my new boyfriend, Ron . . ." But my mother, who'd been so traumatized by my outlandish plans in the past, appeared to have had recently created some temporarily effective, though faulty, defense mechanism that afforded her a few moments to bask in

a serene state of denial before finally accepting the truth later on. Without blinking, she proceeded to tell me about condo and apartment options.

Meanwhile, I scoured Craigslist ads for used vans in the Raleigh-Durham area. I was amazed with not only how many vans were for sale, but also how reasonably priced they all were.

I figured people were trying to sell their excess vehicles because just a few months before, in September 2008, several megabanks collapsed, causing a historic financial meltdown. As a result, there were massive layoffs, home foreclosures, and an unprecedented number of bankruptcies.

America was in a state of panic, and while I worried that my parents might get laid off, I couldn't have cared less about this "Great Recession." I was out of debt, I didn't have any valuable possessions that could be repossessed, and I couldn't go bankrupt because I pretty much already was. I had no assets, no health insurance, nothing. As the rest of the country ran around flailing their arms and screaming in their underwear, I watched half sympathetic, half amused. When you've got nothing, I guess you've got nothing to lose.

Among my options were a red 1989 Chevrolet G20 NASCAR-edition Sportvan for $1,400, a 1994 Chevrolet Astro work van for $750, and a 1997 Dodge Conversion van for $1,700, which was – the ad promised – "an unbeatable road trip machine" that, unfortunately, "needs some repairs."

My needs were few. I wanted a van that was big, under $2,000, in reasonably decent condition, and had never been smoked in (because of all the things I was bound to smell of, I didn't want cigarette smoke to be one of them). And then I found a 1994 Ford E-150 Econoline, a low-top conversion van, listed for $1,500. The ad read:

1994 Ford Van, Runs and Drives Great. Color is Burg. over Black. It has Reclining rear seat T.V and VCR. It has a AM/FM/ CD Changer .The van has appx 119k miles. Everything works. It has the 5.0 V8 , so the gas milage is pretty good for a van. Call with any questions. The price is firm!

Firm. I liked the sound of that. This guy sounded like a straight shooter who didn't mess around with haggling and all that funny business. I e-mailed him, asked if anyone had smoked in the van, and when he responded that no one had, I knew it was only a matter of time before I'd be behind the wheel of my new home.

I knew I needed a temporary "base" in Durham to shop for a van and get situated, so I put up an ad on Craigslist seeking some kind soul to lend me his or her couch for a couple of nights.

My first response was from a local named Kenneth. "hello," said Kenneth. "iam 10 min from your school you sleep on my couch i only have a couch sleep on or sleep in my water bed with me and my wife lol."

I didn't know whether to be more disturbed by the grammatical sacrilege of the English language or the invitation to what could have been an Appalachian-style ménage à trois. I declined the offer after I got another response from Marietta, a Jamaican home health care provider. She picked me up from the airport, and on the drive to her place, I told her about Kenneth's response. She laughed heartily, the laugh starting in her belly and exploding out of her mouth. She said I could stay at her place until I found "proper housing." For whatever reason, I couldn't bring myself to tell her my actual plans. I let her believe that I was in the market for an apartment near campus, hiding the whole van experiment thing, half because I was embarrassed and half because I thought she might think I was insane. Nor did I want Marietta channeling the spirit of my mother, saying (except with a Jamaican accent), "No, mister. You're going to get yourself an apartment."

The day after I arrived in Durham, I took a bus from Marietta's home to a used car dealership called John's Motors south of Raleigh, where the Econoline was for sale. Once at John's, I looked over the lot, scanning the rows of sedans, trucks, and SUVs in search of the Ford Econoline I'd found advertised on Craigslist.

And there it was. A gleaming giant coated with a burgundy sheen, the sun turning its black-tinted windows a blinding white. It looked out of place among the shiny, spotless SUVs, whose bumpers were proudly faced away, as if exhibiting a juvenile disdain for their ponderous, premillennial elder. It was bigger than I thought it would be: a man among boys, a van among toys, a venerable circus elephant whose distended underbelly hung vulnerably low – so low that I wondered if it would scrape its undercarriage when climbing up and over speed bumps.

I approached it with slow, cautious steps, extended my hand, and pressed my palm against its hood as if to feel for a pulse, feeling instead something akin to the voltage exchanged between lovers upon first brushing shoulders. I circled it, lovingly rolled fingertips over dents and chipped paint, took a step back, and admired its burgundy to black "fade" – a color scheme that had fallen out of fashion long ago but was more popular (though still considered tasteless and vulgar) in previous decades.

The Econoline, upon first view, was everything I hoped it would be. It was big, it was beautiful, and best of all, it was $1,500. I was mesmerized.

John was your typical used car salesman. He was a large Italian man who wore dark pants and a black silk dress shirt with gold jewelry cinched around his meaty fingers and wrists. His deceit was poorly hidden behind a thin coating of charm. I knew what I was getting into when he brandished his folksy grin and placed his paw on my shoulder as we shook hands.

I was actually buying the van from a guy named Dennis, who put the ad up, but for whatever reason Dennis wanted me to pick it up at John's used lot. Making things even weirder was their insistence that I pay in cash.

"Looks like ya got a good deal here, buddy," John said, patting the dash on our test drive. "I fixed it up real nice for ya. I even put on a new pair of wipers."

It was true. The wipers were new, but I'd later learn that they only had a couple of settings. On slow, I could count on them to wipe every couple of minutes. I was reluctant to put them on fast because the arms waved so rapidly they created a sort of mirage, granting me an exclusive window into what looked like a wet and blurry parallel universe.

While the ad promised that it "drives great" (which was more or less true), it had more than enough problems to justify its $1,500 price tag: One of the two side doors wouldn't open, there were large patches on the windows where the tint had peeled off, and there were enough dents and scrapes and bruises to have sent unluckier vehicles to be crunched under giant tires at monster truck rallies. Worst of all, the tires were bald – so bald that, later on, when I went to buy a part at Sears, a couple of mechanics doubled over in laughter when I asked them if they thought the tires would pass inspection.

"I'll take it," I told John anyway, unable to hold back my grin. Despite its deformities, it was love at first sight. Plus, I had little choice. My classes were going to start in a couple of days, and I didn't want to overstay my welcome at Marietta's. But when we got down to the business of paying, John tacked on a ludicrous $200 "documentation fee."

"Dennis didn't say anything about this on the phone . . ." I said, heartbroken, knowing that this was precious food money. I'd calculated that I had just enough money to keep me afloat for a couple of weeks – just long enough for me to land a job.

"That's the deal, buddy. You can take it or leave it."

I had no choice. I gave him the $200 and signed my name on the papers with a hard, angry flourish. But as flustered as I was with John, I couldn't have been more thrilled about moving into my new home.

I got behind the wheel and started the ignition. There was a grumble, a cough, then a smooth and steady mechanical growl. I turned out of the lot and headed north toward Duke.

—Day Two of Vandwelling Experiment—

15

....

RENOVATION

SAVINGS: $1,617

WHEN WE THINK ABOUT someone who lives in a van, we probably think of pop-culture losers who had to resort to desperate measures in troubled times – losers like Uncle Rico from *Napoleon Dynamite,* or *Saturday Night Live*'s Chris Farley, who, playing the motivational speaker Matt Foley, would famously exclaim, "I live in a van down by the river!" before crashing through a coffee table. Or maybe we think of the once ubiquitous inhabitants of multicolored VW buses, who'd welcome strangers with complimentary coke lines and invitations into writhing, hairy-bodied back-seat orgies. In the first decade of the twenty-first century, living in a van was for failures and pedophiles, not ascetics and adventurers.

But it wasn't always this way. Americans have been living in vehicles for centuries. Technically, the first American mobile home was the horse-drawn Conestoga wagon that transported and housed pioneers migrating west in the eighteenth and nineteenth centuries. Later, with the advent of the automobile, people almost instantly recognized that cars could also function as homes. The idea took off in 1935 when the house trailer underwent mass production. This new contraption – coupled with the Great Depression – made living in a vehicle an attractive alternative to Hoovervilles. In December 1936, the *New York Times* announced that "We are rapidly becoming a nation on wheels . . . [M]ore families will take to the road, making an important proportion of our people into wandering gypsies." A year later, in February 1937, the *Times* estimated that there'd be two million people leading "a gypsy life." *Harper's* prophesized that the mobile home "will eventually change our architecture, our morals, our laws, our industrial system, and our system of taxation." *Fortune* described the mobile home as "the most promising form of instant low-cost housing since tents and caves and hollow trees." And in 1952, *Newsweek* declared that "the sixth largest city in the United States is on wheels."[1]

But these were mostly mobile house trailers. The van itself didn't enter the scene until 1950 – a pivotal year in vehicle-dwelling history – when the German company Volkswagen manufactured the Type 2 bus, which would become the prototypical "hippie van" for American consumers for the next several decades. The converted vans were complete with seats that folded into beds, birch interior panels, cabinetry, a sink, an icebox, water storage, curtains, an electrical hookup, as well as a large awning that connected to the van's exterior. Americans – typically servicemen – would pick up their VW in Germany, drive it across Europe, and ship it home.

As the van gained popularity in the United States, in 1961 Ford decided to introduce its Econoline model. (Dodge and GM would quickly follow suit with versions of their own.) During

---

1   Credit goes to Michael Aaron Rockland's book *Homes on Wheels* and David A. Thornburg's book *Galloping Bungalows* for their admirable work on the history of vehicle-dwelling.

the late 1960s and 1970s, the vandwelling demographic shifted from families who went on camping trips to hippies who saw the van as a small price to pay for a lot of freedom.

The "vanning" culture exploded, and California was at the movement's epicenter,[2] with 250 converting factories springing up just in Los Angeles alone. "A new form of nomadness is sweeping the country," *Time* magazine said. Groups of vanners began staging large, Woodstock-style gatherings across the country, sometimes with as many as 6,300 vans and 30,000 people. Drugs, alcohol, and sex, quite naturally, were a large part of the festivities, as were topless (and some bottomless) competitions. Terry Cook, author of *Vans and the Truckin' Life*, describes the typical vanning scene:

> Eventually you meet some other guys with vans. You hear about the local van club one way or another and get invited to a meeting. It is like Dorothy when she opened the door of the house that the tornado blew away to Oz and found herself in Munchkinland. Suddenly you meet new friends with similar interests. These Munchkins are into vans, but they are also into other stuff you're into, like rock music or partyin' or goin' places and meetin' chicks — beaver patrol.

As the van grew in popularity, it made more and more appearances in pop culture, like on *The A-Team*, where the four heroes embark on rowdy adventures in their GMC. In *Scooby-Doo*, mystery solvers Scooby, Fred, Shaggy, Daphne, and Velma operate out of a floral multicolored van called the "Mystery Machine." The bad guys in *Starsky & Hutch* often ride in vans. And in the 1977 movie *The Van*, a young man named Bobby tries

---

2 The term "vanning" refers to a culture in which people renovate their vans and camp together at large gatherings, which is different from "vandwelling," which refers to a culture in which people live in their vehicles on a permanent or semipermanent basis.

to lure young girls into his van, which he calls his "fucktruck" and "fuckmobile."

Cook ends his book optimistically: "The movement is big and keeps on getting bigger because the van is a true expression of America's romance with, and now *on,* wheels . . . vans are here to stay."

But they weren't.

While vans never really went away, the vanning and vandwelling cultures faded. Vans became known less for "beaver patrol" and more for picking up your kids at soccer practice. Now, only 2 percent of mobile homes are mobile (only using their wheels to get from the factory to the trailer park), and today's vanning groups pale in comparison to the legendary thousand-strong extravaganzas in vanning's glory days. In the 1980s and 1990s, living in your van – perhaps because of the relative stability of the economy and the end of the counterculture movement – was no longer popular. The days of using your van as a "fucktruck" were long gone.

But in the age of the Internet, there has been a new vandwelling surge. In 2002, the Yahoo! online group VanDwellers was created, which by August 2012 was a watering hole for more than eight thousand members, some of whom hold intimate gatherings in places like the desert near Quartzsite, Arizona. Their ranks increased drastically after the bank crashes and subsequent home foreclosures and have been growing steadily ever since. In November 2011, the *New York Times Magazine* reported that while over a million homes will be repossessed, van sales are up 24 percent. "Living in a van or 'vandwelling,'" the article read, "is now fashionable."

Despite the surge, vandwelling is still just a subculture of a subculture, and mainstream America still considers living in a van as a "creepy" or, at best, an "eccentric" thing to do.

But living in a van just made sense. The average American house, according to author Graham Hill, contains 2,169 square feet – double what it was in the 1950s. Mine would have 60.

Most would agree that it would do us a lot of good if we took up less space, used fewer resources, and wasted less money. Why not live in a van?

I figured that living in my van and adopting a spartan lifestyle would help me graduate debt-free. But there was more to it than mere financial necessity. I also thought that these next few cash-strapped months might give me an opportunity to experiment with a new way of living. If just for a few months of my life, I'd pare down my possessions and strip my life of all unnecessaries. I'd polish my intellect at school but sand down my body to a fine grain with a tough life in the van. I'd embrace a bare-bones, uncluttered simplicity – a voluntary poverty. And if I came to learn that I had been dependent on some pleasure or comfort of a vain nature that did not contribute to my subsistence, then I knew I'd leave Duke having learned at least one thing of value.

But I also hoped – from this upholstered hermitage – to look upon the world around me from a novel vantage point and see my country with an enlightened pair of eyes. I'd be a monk, a hermit, a recluse: within society, yet completely separated from it. I'd hunch over my books and papers and befriend ancient thinkers, never concerned with the neighborhood of man all around me. I didn't need them, and they didn't need me.

Or that was the plan, at least.

Unlike most any other graduate program, the Duke liberal studies program covered all the liberal arts: history, philosophy, literature, anthropology, social sciences, everything. It was a dream program – except for it would cost me.

The tuition for the degree was $11,000, which most would agree isn't bad for a graduate degree nowadays, but because I had no job and now, after buying the van, only $1,600 in the bank – which somehow had to get me through the next four months – affording further education would be (as higher education is for most American students) practically impossible. If I did things in a conventional manner by paying the typical costs (apartment rent, utilities, food, transportation, tuition), I knew there was no way I'd make it to the end of the semester without having to take out loans.

Most other students in a similar predicament might have either taken out loans, kept working to save money, or enrolled in a graduate program that paid their tuition. Or they might have just decided to not go to school. These were all sensible and logical solutions, but postponing my reentrance was unthinkable.

A couple of years before, I'd tried to get into ten Ph.D. and creative writing programs that offered assistantships, which would have provided me with free tuition and a modest stipend. All those applications had been rejected, as were the three scholarships I filled out after being accepted by Duke. While I'd hoped to sneak my way around college tuition, I'd met warty, mustachioed border patrolmen on all the conventional roads to an affordable education.

No big deal, I thought. The van, I justified to myself, would be a fun, if just temporary, fix. Because I could return to my well-paying job with the Park Service in the summer, I knew that all I had to do was live in a van for this first semester. According to my calculations, after another summer with the Park Service, I would save up enough money to allow me to pay for the rest of my graduate education. I'd suffer in the van for the first semester, then sell it and upgrade to an apartment during the next fall semester. And still live debt-free.

If Henry Thoreau was my philosophical mentor, Bob Wells was my practical adviser. I'd been reading his how-to vandwelling website, CheapRVLiving.com, for months, learning everything I could. On his site, Wells gives advice on everything you need to know about living in a van: how to pick a van, how to ensure "stealth," how to install solar panels, even how to go to the bathroom.

Bob hadn't always been such an expert. He started vandwelling in 1995, when his wife of thirteen years divorced him and kicked him out of the house. Because he had little money and only a low-wage job at a supermarket, he bought a $1,500 Chevy box van to be both his home and means of transportation. The first night he slept in his supermarket's parking lot, Wells cried himself to sleep.

Over the next few months, he made a number of improvements. He brought in an ice cooler, a propane cooking stove, a TV, and even a recliner. He lined the walls with insulation, made shelves for food and supplies, and acquired a propane heater. He slept on a homemade bunk bed, used public restrooms, rented a P.O. box, and started going to a gym where he'd take showers. When he had to go to the bathroom at night, he'd use a "pee pot" that he'd empty out at red lights on drives around town.

As Wells grew more comfortable in his van, not only did he stop crying himself to sleep, but he also began embracing the simple life in ways he never would have expected. He had fewer bills and more time to himself. He cut down his hours at work and stopped buying useless things because he didn't have any extra space to store them.

He ended up retiring from his job, spending most of his time traveling, occasionally working as a seasonal campground host to make money. To share his love of vandwelling and help people transition into their vans, he created his website, which is now considered by many vandwellers to be the authoritative guide on how to convert your van and renovate your life.

If there was anyone who could tell me if secretly living in a van on a college campus was possible, it would be Wells. I e-mailed him and eagerly waited for his response.

To: Ken Ilgunas
From: Bob Wells
Date: December 12, 2008

Subject: Re: question from vandwelling disciple

Hi Ken, the main potential problem is campus security. Generally, if they find out you are living in your van they won't like
it. I can't really add more since I have never done it. My sug-
gestion is to join the vandwellers group on Yahoo and ask the
question there. I know there are people in the group who have
done it.

Sail on,

Bob

Hmm. Not exactly the sage advice I was hoping for.

I scoured the Internet for more guidance and found the
Yahoo! message board called "VanDwellers: Live in your Van
2." There were thousands of members on the site and plenty of
useful tips, but no one on there had secretly lived in a van on
a college campus or could reassure me that I might get away
with it. While I could only presume others had lived in similar
fashion to afford their educations, I had no precedent to follow.
I felt unsettled yet invigorated, as if I were venturing out into
unexplored territory without a map or guide.

Oh, the freedom! There are few sensations as liberating as being
in the driver's seat of your own vehicle. Yes, there'll be bills. Yes,
you now *own* something that'll weigh you down. Yes, you're
sorta destroying the environment. And yes, this was the sort of
jalopy that might have given desperate Depression-era Okies
cause to consider sticking out the Dust Bowl rather than testing
the vehicle's reliability en route to a new life in California. But
all reservations aside, the freedom the vehicle lends its owner
feels – at least at first – worth all those costs. I'd no longer have
to hitch rides and depend on other people for transportation.
I'd no longer have to travel according to the schedules of buses,
planes, and trains.

As I drove north to Duke after buying the van at John's, I had
an overexcited, caffeinated, "I'm going to pee out of my ass-

hole" feeling in my stomach. I had so many questions and so few answers. I had a van, but I had no idea where I was going to park it. I had bills, but no job. I had a home, but no idea how to live in it. I had to go to the bathroom, but didn't know where to go and didn't want to follow Bob's advice.

My first class was in two days, so I had to transform the van into a home as quickly as I could.

My first order of business was to get rid of the van's two middle pilot chairs. I figured I'd need some room for cooking and getting dressed and just living in general, so I put up an ad on Craigslist seeking someone who could store my seats. I offered $30 and within an hour I had multiple offers.

After dropping off the chairs at a local's home (who, thankfully, didn't ask any questions about why I needed extra space), I drove the van to the nearest Walmart to begin renovating. I jumped into the back of the van and visualized what I wanted my home to look like. I had fantasies of installing solar panels on the roof, a periscope so I could see all around me, a wall of shelves, houseplants, a hammock! But I had very little time and only a bit of money. I could only hope to make the van comfortable.

On closer inspection, I discovered a whole bunch of stuff that I'd failed to notice on my initial tour at John's. On the floor in between the two front seats were a mounted TV and VCR. And under the passenger seat was a twelve-disc CD changer with country star Alan Jackson's best-of compilation *Super Hits*. All the windows were tinted, and they all had beige pull-down blinds (except for the windshield, of course). The doors and windows were automatic. On the van's rear was a spare tire and hitch, and there were a half dozen lights on the ceiling that could be clicked on.

Just when I thought I had uncovered all the van's secrets, I found a mysterious button inside on the driver's-side wall at the van's midpoint. I pushed it and the backseat grumbled, vibrated, and – much to my jubilation – began slowly and suggestively folding down into a bed.

With the bed down and the middle chairs gone, I saw that I had more than enough living space – probably enough room to stretch and do push-ups if I wished. I took out all the stuff I'd brought and laid it across the bed before neatly packing it away.

Miscellaneous	Clothes / Shoes	Camping Gear
Backpack	Suitcase full of clothes	Large backpack
Books from home	Five dress shirts	Ultra-light backpacking stove (for cooking in van)
Journal	Two pairs of dress pants	Water filter
Laptop and laptop bag	Slippers for shower	Matches / lighters
Travel bag with toiletries: shampoo, soap, shaving supplies, toothbrush, toothpaste, etc.	Shoes, sneakers, hiking boots	Sleeping bag (rated to -20°F)
Notebooks	Two towels	Tent
Multi-tool / hunting knife / cooking knife	Gym clothes	Tarp
Sewing kit	Expedition-rated thermal underwear	Water bottle
Foldable laundry basket		

I had with me nearly every useful thing I'd accumulated over the past two and a half years: a –20°F rated sleeping bag that I'd bought from a friend for $40; a set of expedition-rated thermal underwear my mom had bought me when I lived in the arctic for a winter; an MSR ultralight propane backpacking stove that I'd bought for $85 before I journeyed across the continent; and a twelve-inch-long hunting knife for protection that I'd bought on the voyage from a hunter in Ontario, Canada. I also had a heavy, stout knife for cooking and a multi-tool that would be my can opener, screwdriver, and toenail clippers. I considered all of the above to be almost equivalent to "necessities."

I went into Walmart and came back to the van with a cart full of supplies. I screwed five hooks into the ceiling behind the two front seats and cut tiny holes into a huge black cloth, so I could hang the cloth on the hooks. Between the black sheet, tinted windows, and blinds, I figured I now had "stealth," as the vandwellers called it. (While I thought a big, creepy van with blinds and a black sheet behind the front seats might look suspicious, all I could do was hope that no one would ever even think that someone was living in his vehicle at a place like Duke.)

I screwed two large coat hooks into opposing sides of the van's interior, where the middle pilot chairs used to be. One would serve as a coat hook and on the other I'd hang my dress shirts and pants.

I neatly folded the rest of my clothes into my suitcase and pushed it under the bed. At Walmart, I bought a plastic three-drawer storage container, which I would fill with bulk food and miscellaneous items. The two-and-a-half-foot-tall storage container would be a perfect counter for cooking my meals on my backpacking stove.

When I took a careful turn out of the Walmart parking lot, all my stuff gushed out of the drawers, so I went back in and picked up bungee cords to securely hold my drawers in place. At the Salvation Army, I bought some sky-blue bed linens, a white blanket, a small wastebasket, a pot, and a pan – all for $10.

I knew that food was probably going to be my biggest expense after tuition and the van, so I planned on doing all of my own cooking. I went to a Kroger supermarket and loaded up on cheap bulk food. I bought a four-pound box of powdered milk; bags of beans, rice, and spaghetti noodles; several canisters of oatmeal; an eighty-ounce container of crunchy peanut butter; and a dozen large boxes of cereal. I also picked up a couple of small eight-ounce isobutane canisters for $19.

On top of everything I just bought, I'd soon pay my first car insurance and cell phone bills. I hated the idea of paying these, as I hardly considered them "needs," but I knew from past experience that I would need a phone for job purposes. I signed up for the cheapest plan I could find ($37 a month), which gave me 200 minutes a month (no texting) and free weekends. Car insurance also seemed like it would be an unnecessary drain on my account, but because I wouldn't be able to get license plates without it, I bought the cheapest plan I could find for $47 a month.

I needed an address and a mailbox, so I got a P.O. box at the campus for $21 a semester, as well as a membership at the campus gym for $34 a semester so I had somewhere to take showers. I bought used books for my two courses for $95. The last big expense was a campus parking permit for $182, which would pay for parking next fall as well. (I figured I needed a parking permit because I had no idea where I could routinely park in the city.)

When I was an undergrad, I spent my money carelessly, and rarely was I aware of how much I had in my account. Nor did I know how much debt I was in or how much debt I was going to have. I would do things differently this time. This time I'd be meticulous. I'd keep track of every penny. If I spent money, I'd write it down. I'd keep all my receipts. And each night I'd review the balance to make sure everything was in order.

The following is what I bought during my first week at Duke:

Van	$1,500
Documentation fee	$200
Van plates and processing fees	$128
Van renovation materials:	($45.85 overall)
• Screw hooks	$1. 59
• Bungees	$3.00
• Linens, blanket, waste basket, pot and pan	$10.33
• Thumbtacks	$0.88
• Coat hooks	$1.08
• Fabric	$9.00
• Three-drawer plastic storage container	$19.97
Gas	$50.63 (30.888 gallons at $1.639 a gallon)
Seat storage	$30
Isobutane fuel	$19.05
Food	$107.33
Automobile insurance/month	$46.43
Cell phone/month	$37.30
School costs:	$372 (overall)
• P.O. Box	$21
• Transcript fee	$40
• Gym (for semester)	$34
• Parking permit (for year)	$182
• Books (all bought used)	$95
Total:	**$2536.59**

When all was said and done, I looked aghast at my balance: in one week, I'd spent $2,536.59, which was an alarming 72 percent of my savings. I only had $981 left, and I hadn't even begun to pay the tuition, which was $2,178 for the semester. At this rate, I knew I wouldn't get past my second week of school, let alone the next two and a half years without having to go back into debt.

I reminded myself, though, that most of these were one-time purchases, and that from here on out I'd only have to worry about tuition, food, car insurance, and cell phone bills. The stuff I bought today, I justified, would help me save tomorrow.

It was true: For one, I had no rent or mortgage bills to pay. The cheapest apartments in Durham were about $450 a month (utilities not included) and the Duke underclassmen (who are forced to live on campus for their first three years) had to pay far more.

The van made perfect sense. It would cost me nothing. Literally nothing. While I paid $1,700 for it, I figured I could get that or almost as much as that back when I'd sell later.

The van was just one way of cutting back. I figured I'd do without other things that were running millions of Americans into debt, like health care, entertainment, clothes, and transportation. For transportation, I'd walk and take the bus. If I needed new clothes, I'd sew up tears or buy used stuff at the Salvation Army (where a good shirt or pair of pants is rarely more than $3). I'd entertain myself with books and movies that I could get for free at the library, and while I always felt ill at ease without health insurance (which I hadn't had for much of the last three years), there was no way I could afford it. To stay out of debt, I'd have to stay healthy.

Tuition was the biggest threat to my experiment. Normally, a sentence that has "Duke" and "affordable" in between a pair of periods would give one cause for incredulous laughter. But Duke's liberal studies program was affordable if you could demonstrate financial need (which I could). My program reduced my tuition from $3,131 to $1,089 a course (which would cost me, in total, about $11,000 for the degree). Because I was out of academic shape and because I needed to devote a significant portion of my schedule to a part-time job, I decided to take just two courses instead of the standard full-time three-course load. I'd have to pay $2,178 for the semester in four installments – one payment of $544 each month, the first of which was due in four weeks.

I spent the rest of the day in the Walmart parking lot, putting the final touches on the van and reading *Moll Flanders*, which was assigned reading for my first class. I pulled down all the blinds, slung up my giant black cloth behind the front seats, put on a long-sleeved T-shirt and sweatpants, and squirmed into my sleeping bag on the backseat, which I'd turned into a bed with the push of a button.

Between being thrilled about the prospect of attending classes and preoccupied with the Walmart security guard who strolled past my van every thirty minutes, I could hardly sleep.

I had my van and it was ready for vandwelling, but I still had so much to figure out. I needed to find a job, learn how to cook in the van, find a place to wash, and pass my classes. And, of course, I'd have to do it all without going back into debt.

Finally lulled by the hum of distant traffic, I fell into a deep slumber from which I didn't wake until the early hours of the morning.

—Day Seven of Vandwelling Experiment—

16

....

ACCLIMATIZATION

SAVINGS: $981

BEFORE MY EXPERIMENT BEGAN, I knew I had the personality for vandwelling. Over the course of my journey to get out of debt, I'd developed a penchant for rugged living, a comfort with tight quarters, a sixth sense for cheapness, and a tolerance for squalor that was, well, (I hate to brag) *unequaled*. Not only that, but I knew I had the physical constitution for it, too: I was blessed with a high tolerance for cold temperatures, practically no sense of smell, and a bladder (I hate to brag) the size of an adolescent's football.

But I wasn't your ordinary vandweller. Unlike famous vandweller Bob Wells and his vandwelling acolytes, I couldn't just pack up and skip town whenever I wanted. I, rather, would have to live a moored existence on campus. And I knew that if I was going to get through college debt-free, I'd have to keep the van and my experiment a secret.

While I'd been living in remote and rural places for the previous couple of years, and had been, during that time, completely cut off from college culture, I still knew the student mind well enough to know that if one of them discovered my secret, a story of a dude living in his van on campus would probably become the sort of mindless gossip that would give students something to procrastinate with for a day or two. Perhaps it would begin when a fellow student – in the wrong place at the wrong time – took a blurred picture of me exiting the van. He'd put it up on Facebook and it would go viral over the Internet, spreading via e-mail and Twitter at terrifying, lightning-quick, real-time speeds. And as soon as campus administration caught wind of it, some frantic pantsuited woman would be sprinting down an office hallway, waving a sheaf of papers over her head, worried that a story of a cash-strapped grad living in his van would become a PR nightmare.

In order to graduate debt-free, I'd have to lie. There was no way around it. This was all very troublesome because the art of lying is an art I know nothing about. If I were in a heist movie, I'd be the guy on the team who – during the opening scene's first certain-to-fail robbery, buckles under pressure at the last minute, fucking everything up, leading to some doomed car chase where I – to the audience's relief – am brutally murdered by cops as the team's only victim, allowing the gang to hire better, more competent men for future jobs.

On my first day at Duke, I went to the parking and transportation services office to buy a parking permit. I walked in with sweaty palms and shifty eyes, worried that the place would be crawling with campus security guards keeping an eye out for students up to no good. When I asked an administrator for a form, beads of sweat gathered on my forehead and, when they reached the appropriate size, streamed down the curves of my face like luges down hill slopes. Part of me wanted – right then and there – to fall to my knees, confess my plans, and beg for mercy.

The whole time, I thought, *Ken, you have no idea what the hell you're doing.* It was true. I'd never been to Duke, and I hardly knew anything about the school. And I certainly didn't know anything about whatever parking lot they'd assign me to. Beforehand, just to make sure, I read over Duke's parking regulations and was pleased to discover that there were no anti-vehicle-dwelling laws. I presumed that the absence of such a law had nothing to do with their lenient policy of allowing students to save money and experiment with housing. Rather, I figured such a problem simply had never come up. They probably had no reason to have a regulation banning vandwellers.

My biggest worry was the lot they'd assign me to. I learned on their parking website that as a first-year grad student I could be assigned to any number of lots. And I had absolutely no say about which one I'd get. As far as I knew, I could be sent to a busy, packed lot within eyeshot of an office window where men and women in business attire – if they learned of me – would make smart remarks about my sanity, love life, and living conditions. ("Cheryl, look! He's peeing in the sewer drain!")

Despite all my misgivings, I was blindly confident and stubbornly determined. I'd make whatever parking lot they gave me work, I told myself. And I'd greet each tribulation as a noble challenge and never as an unwanted adversity.

Registration was easy enough, except that to get a parking permit I had to put down a local address (which I of course didn't have). I had my laptop with me, so I flipped it open, checked out Google Maps, looked up Marietta's address, filled it in on my application, and hoped that no one ever double-checked that sort of stuff.

When Thoreau first saw Walden Pond as a young boy, he was tantalized. To him, it was a "recess among the pines where almost sunshine and shadow were the only inhabitants." It became, as he described it, "one of the most ancient scenes stamped on the tablets of my memory."

When I first cast eyes on the parking lot Duke gave me – the Mill lot – I had none of the flowery thoughts Thoreau had upon seeing his future home.

I wasn't sure if I should even bother with my experiment. The lot looked terrible. I had hoped for some quiet, rarely visited lot that would give me some semblance of privacy. This . . . this was the exact opposite.

The Mill lot, for starters, isn't even located on Duke property. Nor is it anywhere near the main campus. (While it is only a quarter mile away from Duke's East Campus, the East Campus, unfortunately, is a mere island colony of Duke's much larger, busier, more happening West Campus.) I didn't mind a little walk to campus, but that wasn't the problem. The problem was that the Mill lot is located on Ninth Street, a busy, bustling mini metropolis of bars, cafés, restaurants, and new age shops where bums, bikers, students, and yuppies commingle.

I was surrounded by buildings and people. How could I keep myself a secret *here?*

In front of my parking spot, about twenty yards away, was an old three-story redbrick tobacco mill that had been refurbished into a stylish apartment complex where mostly upperclassmen at Duke lived. Behind me was a student bar called George's Garage. To one side of me was Ninth Street and on the other was a large, empty grass field where apartment dwellers walked their dogs.

Because my permit allowed me to park in whatever Duke parking lot I wanted after 5 P.M., I wondered if I should just drive the van to a new lot each night. But to get through the semester debt-free, I knew I needed to save gas money and avoid wear and tear on the van. With that in mind, I decided that I ought to try to adapt to life in the Mill lot. Plus, I figured that campus security would get suspicious if they saw my van in a new lot each night. This way, by staying in the Mill lot, I hoped campus security would presume that I was a student living in the adjacent apartment complex.

I parked the van in a section of the parking lot that was relatively empty, pulled down my shades, hung up my black sheet behind the two front seats, and set out to walk the quarter mile to Duke. This would be the first time I'd lay eyes on my new school.

Duke University, originally called Trinity College, was founded by Methodists and Quakers in 1838. It expanded when the Dukes, a wealthy family in the tobacco industry, gave generous donations to the school with stipulations that women be put on equal footing with men. What was once a tiny, inconsequential college is now home to 13,500 students on 8,610 acres of campus.

I walked from East Campus to West Campus, zigzagging around bronze statues of the Dukes, strolling under stone archways, and wandering through a labyrinth of hallways in the medical center.

To say the least, I was impressed.

Duke looks every bit like a college on a university brochure: it is a leafy green village of redbrick mansions and Gothic stone monasteries, populated with an unreasonable plentitude of good-looking young people. Beneath towering hundred-year-old oaks are crisply shaven lawns that in warmer weather would be spotted with sunbathing girls and Frisbee-flinging guys. In the middle of campus is a tranquil fifty-five-acre garden that attracts visitors from across the region. The university features a thirty-four-thousand-seat football stadium, a forest of 7,000 acres, and Duke Chapel – the centerpiece of campus – that stands more than 200 feet high and houses a bell weighing 11,200 pounds, three organs with 12,633 pipes, and a crypt with the remains of three Duke presidents. Amid the smell of refinement and riches, though, was a delightfully plebeian stench hovering above a tent village called Krzyzewskiville (or "K-ville"), set up on the lawn near the basketball stadium. There, hundreds of students wait in line for months on end with hopes of being awarded the coveted tickets to the Duke men's basketball team's match against their archrivals, the University of North Carolina Tar Heels.

When I arrived, Duke was ranked as the tenth-best school in the nation by *U.S. News & World Report* and fourteenth in the world by *Times Higher Education–QS World University*. With a $5.7 billion endowment, Duke is the fifteenth-wealthiest university in the United States. And it's one of the most difficult schools to get into, accepting only 16 percent of the students who applied in 2011.

While there are various ethnicities at Duke, a significant portion of the student body hails from the wealthy. It's the sort of place where, in the glass-ceilinged Von der Heyden Pavilion coffeehouse, you could expect to overhear frat brothers Brant and Chaz talking about how they'd spent their winter break at their fathers' châteaus in the Pyrenees. This was all so different from Buffalo – my depressed industrial hometown, where the people are blue-collar and the sports teams, perennial (though lovable) losers. While Buffalo had been made rusty by raw deals, Duke scintillated with success.

Given my modest academic record, I assumed Duke's admissions department had made some flabbergasting, "someone oughta get fired" error in welcoming me into their elite ranks. I did well in college, but not so well to think I'd ever end up in a place like Duke. I got in, though, presumably because I was a perfect fit for their liberal studies program (which, conveniently, was the least selective of all of Duke's grad school programs). The selection committee, according to their website, is less interested in grades and more interested in applicants with "postgraduate experiences and recent accomplishments," who have "broad intellectual interests," and who possess "the capacity and energy for learning, writing, and discussing ideas." I didn't have the grades and accolades, but I had all that other stuff.

The liberal studies program appealed to me because I had no single academic interest. I loved all the disciplines: literature, history, science, and even a little bit of math, so it would have seemed unusually restrictive to be forced to narrow my studies to one field. I didn't want to study anything I wasn't curious or passionate about. And the last thing I wanted was to become a "specialist" – an expert at one thing but useless at everything else. (I preferred to remain useless at everything.) While specialized study often makes new scholarly discoveries possible, I wasn't so interested in writing academic papers that only half a dozen fellow experts would sleepily read, using big, boring words like "empirical," "paradigm," and "ontological." I wanted to focus on bigger-picture stuff. I wanted – with the help of professors and classmates and the great texts – to learn how to live the best life possible. I wanted to be an "artist of life": someone who knows how to live, how to die, how to be happy and valued and necessary and good. I didn't want a degree or a career. I didn't want a "marketable skill" – I could learn one later if I wished. It seemed like now – before amassing things and obligations and a career – was the time to figure the important stuff out.

The liberal studies program offered courses that were designed to be interdisciplinary, meaning that the course might contain elements of, say, biology, history, and philosophy all nicely wrapped into one. Plus, I was free to enroll in courses from any other graduate or undergraduate department I wished. If something in the English undergraduate department struck my fancy, I could enroll. If I wanted to learn about natural resources law at the Nicholas School of the Environment, I could. This was true academic freedom. And while I wasn't getting paid to go to school by doing research or teaching undergraduate classes like many grads were, I was most definitely free to study whatever I wished – and that, I figured, would be worth the tuition I paid.

As I walked around campus, there was electricity in the air. Students had been recharged by their winter breaks. All the baggy eyes had been smoothed, all the stubbled faces shaven. I was as excited as ever. I couldn't wait to recapture the sensations from my undergrad years: the invigorating classwork, the stimulating conversations, the provoking lectures. I figured I'd make friends with a trio of like-minded classmates, join a few clubs, find a part-time job, and maybe even write for the school newspaper, as I did in Buffalo. Except now, as I walked around the campus, I felt old and out of place. It was as if I had awakened from a slumber in a cryogenic chamber and walked out to a very different and slightly unsettling futuristic world. Almost all the girls wore skintight black leggings for pants, freely advertising what in previous years would have been top secret topographies of their lower halves. Even weirder was their fondness for a hideous style of fur (or fur-like) boots – commonly referred to as "Uggs" – that made the girls look like Oompa-Loompas when paired with the skintight leggings. The men also were curiously attired, donning large black plastic shades and pastel-pink polos tucked into their pants. Half the student body walked around with cell phones held inches from their faces, busily thumbing buttons in midstride.

The two courses that I'd enrolled in were called "The Self in the World" and "Biodiversity in North Carolina." In Self, we would read texts from the seventeenth century to the present – texts like *Moll Flanders, Their Eyes Were Watching God,* and *Mrs. Dalloway* – which we'd use to examine how ideas of individual identity, subjectivity, authenticity, and autonomy have changed. In Biodiversity, we would learn how to identify flora with microscopes and magnifying glasses. And we'd also look at the environment from a broader vantage point: studying land-use history, the conservation movement, and the ecology of natural ecosystems. The course would culminate with a weeklong field trip to a biology research lab in the Appalachian Mountains.

Depending on your tastes, these courses may seem boring or unusual or a waste of time. But I couldn't have been more anxious to delve into my assigned books and stay up late writing papers that would force me to think about stuff I never had cause to think about before. I browsed over the list of books on the syllabi like they were succulent entrées on a menu.

My first Biodiversity class was that night. I was the first to arrive, and I eagerly introduced myself to a young woman my age who walked in next. We had a nice chat at first, but the tone changed dramatically when she asked where I was living — a question whose ubiquity no one appreciates until you're secretly living in a van.

I clasped my trembling hands under the table, looked down at my desk, and muttered, "Off campus," which was true but deliberately misleading.

"Oh yeah? Me, too," she said. "I just moved into an apartment off of West. Whereabouts are you?"

"I'm on Ninth Street," I said with surprising aplomb, having prepared myself for this very conversation.

Disaster was averted after a few more classmates ambled in and I resorted to taciturnity. I realized then that as long as I had to keep the van a secret, my experience here at Duke would be nothing like my memories of college life at Buffalo.

—Day Thirty-five of Vandwelling Experiment—

17

························

ADAPTATION

SAVINGS: $830

I WAS LYING IN BED in the van. It was 2 A.M. on a cold February night. Moments before, a nearby Ninth Street nightclub had turned off the bowel-shaking, thigh-grinding techno music and purged itself of its scantily clad patrons. Hundreds of students briskly walked from the bar and past my van to their apartments. I knew that even the slightest squeak from the bed could give me away, so I stayed perfectly still and covered my face with my sleeping bag to muffle the sound of my breathing.

A drunk male in midstride exclaimed, "Isn't this the van from *The A-Team*?!" — a snarky comment to which his entourage responded with uproarious laughter. Two girls sat on the curb next to the van, one of whom was sobbing sloppily because she just got dumped by her boyfriend. "Sweetie, you don't know it yet, but this is *good* for you," her friend said admirably. "You're so much better than Steve!"

The students continued to flood past the sides of my van.

Someone walked into the rear of the van face-first. Another braced his arm against its side and vomited into the parking spot next to me.

My one recurring thought: *These people are all getting laid tonight.*

I'd been living in my van for four weeks, and the experiment, so far, had been a complete success. I was debt-free, the van was still a secret, and my brain felt like it'd been properly exercised for the first time in years. In my Biodiversity course, I was reading wilderness philosophers: Aldo Leopold, Roderick Nash, William Cronon, and Jack Turner. In Self in the World, I was having stirring debates over Blackboard and doing everything I could to prepare for class discussions. To give myself a creative outlet, I started a blog called *The Spartan Student,* which no one read except for friends from back home (if just out of kindhearted support), but in case some stranger happened upon it, I kept my school and identity a secret. All compartments of my brain were on fire.

Yet my experiment was no peaceful sojourn on Walden Pond. I spent nearly every moment in a state of anxiety about my financial situation. I only had about $800 left from my savings, yet I still had to pay tuition.

I knew I'd have to cut back on all costs until I found work. The only bill that I had any control over at this point was food, so I decided that I'd eat as little as possible—just cereal or oatmeal for breakfast, a banana and peanut butter sandwich for lunch, and a light pasta dish for dinner.

But after the first week on my new diet, my hunger was constant. And my light, meager meals did nothing to calm the gurgling, clawing, "I'm going to put you through a world of pain if you don't feed me" feeling in my gut. After just a week at Duke, upon weighing myself at the gym, I noted how I was already five pounds lighter. While I shaved in front of the mirror in the locker room, I saw how my ribs rubbed against my skin for the first time in years. And while I thought it might be nice to one day admire a set of chiseled, baby-smooth abs, I knew I had to start eating more when I saw a bunny on campus and was tempted to hurl a rock at it so I could devour it raw. There was only one thing on my mind now: *I need money.*

I got a part-time job as a research assistant for a professor in the business school. For six hours a week at $11 an hour, I made copies, fetched library books, and performed other menial office tasks. I also had to scroll through and assemble data from about five thousand businesses on a Microsoft Excel worksheet, which caused red lines to web across my sclera and my vision to blur. But the money wasn't enough. Once, upon walking to the van in the middle of the night, I saw an old pizza box on the lawn – how long it had been there, I wasn't sure. I opened it and saw a few mangled slices. *Has it come to this already?*

My mother, still in denial about my van plan, began to grow suspicious when, in our e-mail correspondence, I repeatedly failed to address her question regarding the whereabouts of my new home. I thought it would be silly to have to sustain a lie like this for the whole semester, so I resolved to mention the van casually, sandwiching my admission between mundane, everyday details, hoping she might think living in a vehicle was a mundane, everyday thing, too. "Hello mom," I wrote. "I've been playing basketball every day. It's been a lot of fun. I've been eating quite well, and I've been sleeping in my van – which is quite spacious. All is well, Ken."

Her response – evidently restrained – communicated to me her most prominent concerns.

To: Ken Ilgunas
From: Sistine Ilgunas
Date: January 28, 2009
Subject: Re: check
how do you clean yourself? Where do you park the van?

To: Sistine Ilgunas
From: Ken Ilgunas
Date: January 29, 2009
Subject: Re: check
Hey mom,

I park the van in the parking lot, silly. I've been taking showers at the gym. Classes are going well – the professors brought cheese and wine. Everyone has been really nice. Too-te-loo.

To: Ken Ilgunas
From: Sistine Ilgunas
Date: January 29, 2009
Subject: Re: check
how many people are in your class. What will u say when someone asks, hey ken, where do you live? how many other students live in their van? just interested.

To: Sistine Ilgunas
From: Ken Ilgunas
Date: January 30, 2009
Subject: Re: check
There are about 16 in my class. When people ask where I live, I say I'm still looking. I doubt few if any other students sleep in their van. Later.

To: Ken Ilgunas
From: Sistine Ilgunas
Date: January 31, 2009
Subject: Re: check
Hi Ken,

You worry me & you know it. Please let me give you some money. If you are so upset about it you can pay me back. Please get an apart. or roommate or something. Your life must be so stressful the way you are living. How do u form friendships when u live the way you do? I feel so sorry for you. You go to this fantastic school & you are living like a homeless person. How do u explain your life to new acquaintances? Dont you have any self worth? You are always welcome to borrow money or have money from me. Why cant u take help from your family? I am always here for you.

  Love, momxxxooo

PS: Do you want me to pay your cell phone bill it came in today? I will.

My poor mother. None of this made any sense to her. And of course I understood why. My mom, as a girl, shared a small apartment with her parents and two siblings in blue-collar North Tonawanda, New York. My dad grew up in a house crowded with seven brothers and a sister in Motherwell, Scotland, an industrial town. My mom was embarrassed that her mom had a chicken coop in the backyard. My dad got fruit for Christmas. They grew up in industrious middle-class families who knew that you could make it by working hard. She and my father had spent their lives working so they could provide better lives for my brother and me. They upgraded from apartment to home, from city to suburb, from middle class to a few echelons higher in the middle class. All that hard work. All that climbing. All that moving up. And all that time and money invested in me, so that I could move up, too. And this is what their son lives in . . .

I am a member of the "career-less generation." Or the "screwed generation." Unlike previous generations, the members of my generation won't get jobs and respectable wages straight out of high school, let alone college. We don't have the means to buy homes and start families in our twenties. We're the first generation in a while who will be less well off and less secure than their parents'. Strangely, I seemed more okay with this than my parents. Not being able to afford an aboveground swimming pool and a kid wasn't some heartbreaking tragedy to me.

For some, it's hard to think that the direction of success is anywhere but up the socioeconomic ladder, especially when success is largely measured by security, comfort, and wealth. But maybe progress can point in funny directions. Must we measure our success by the size of our homes and salaries? What if we got healthier, lived more sustainably, and became more self-reliant, albeit in tighter dwellings and in smaller families? Isn't that success, too?

I thought about my mom's offer to pay for my apartment rent. It was tempting to think about a warm apartment and lavish feasts, but I couldn't accept her offer. I was determined to graduate debt-free, and that meant I couldn't take out loans of any kind, whether they came in the form of gifts from loved ones or food stamps from the government. Perhaps I took the implications of accepting a gift too seriously, but I couldn't stop the alarms from ringing in my head whenever I considered taking a gift. I didn't think of a gift merely as a gift but as a debt with a bow wrapped around it. The exchange may seem harmless, but I knew my accountant within was always diligently at work, carefully recording in his ledger my gifts and loans and debts, none of which could be truly forgotten until they were paid back.

When we accept a gift, I thought, sometimes we don't just acquire a debt but an identity. Taking a gift can be like taking a sizzling-hot brand to the backside. The giver gives us a marking, a bubbling scar that only the brander and branded can see. It's a mark of dependence.

A couple of my fellow rangers at the Park Service told me that every year, after the summer season ended, they'd apply for unemployment checks. They'd apply even though we'd been paid extremely well during the summer — well enough, I thought, to be able to live off our summer savings for the rest of the year if we lived modestly. Yet one ranger told me that every winter he would go on vacations to the Caribbean, and from his lawn chair on the beach, with a piña colada in hand, he'd call his unemployment officer and tell the officer that he'd been looking for jobs but couldn't find any, feigning exasperation about how "times are tough." With the unemployment checks, he could travel wherever he wanted, buy nearly anything he wished, and didn't have to work for eight months of the year. But is freedom from work really freedom? Is our money – no matter how we acquire it – a ticket to freedom?

I thought that if I accepted money from the government or a friend or a family member, I'd be permitting someone to draw the edges of my identity, too. Perhaps this would be an unnatural way to go through life, always being so unreceptive to others' generosity. And I worried my policy would wear on my family relations, but I didn't want to head down a slippery slope, as it's always easier to ask for the second favor than it is the first. Best to abstain from all offers of financial support, I thought. I knew my mother didn't have anything malicious in mind, but I wanted, once and for all, to be on my own, even if it required that I remain cold and hungry.

While waiting for the bus (connecting Duke's East Campus to its West, which is free to students), I noticed an advertisement stapled to one of the benches. It was put there by the neuro-science department, which was seeking study participants to undergo cognitive tests for $10 an hour. I went online and signed up for every test I could. I spent the next couple of weeks taking cognitive tests and also being zapped by electrodes, pricked by needles, and dazed by pharmaceuticals. (Shamelessly, I donated three of my four primary bodily fluids.)

Bringing in additional money always put me more at ease because I knew it was the only thing keeping me from Dumpster diving or, worse, dropping out of school. Yet I still always felt like I was one mishap, one mistake, one doctor's bill, one unforeseen cost from going bankrupt. At one of these studies, I discovered another flier that advertised Duke's MRI studies.

The MRI studies paid $20 an hour to participants who were willing to go into an MRI machine to have their brains scanned. At first, I was reluctant to sign up. I felt like I'd be selling my body for money if I did something like that. It seemed little better than bartering a kidney to pay a gambling debt or turning a trick to make a car payment. But I justified that I'd be helping science, and that I wouldn't be making money for the sake of making money but for the sake of feeding myself.

Even though the consent forms for the MRI tests assured that there were "no ill effects reported from exposure to the magnetism or radio waves," the following sentence warned, "However, it is possible that harmful effects could be recognized in the future." *Harmful effects could be recognized in the future?* What the hell does that mean?

For my first study, I walked to the radiology lab, took off all the metal I had on me, and lay down on the bed. A middle-aged female technologist with a Russian accent strapped a breathing belt around my waist and clamped electrodes around my left index finger with which she'd shock me with low voltage at key moments during the experiment. She hooked up an "eye tracker" on my plastic helmet that would collect data when my pupils dilated. She stuffed plugs into my ears and snugly packed a pair of headphones against the sides of my face.

"You vant anudder blanket?" she asked.

"Oh, no thanks. This one is fine," I said, before she proceeded to tightly bundle me in a second.

"Comfortable?"

"Very."

"Don't fall asleep."

"I'll do my best not to."

"You better not. You don't vant me to come get you."

She rolled me headfirst into the cocoon-tight cylinder. I could feel my arms (which were already tightly pressed against the sides of my torso) rub against the walls of the machine. After being slid in, I began to feel the first tremors of panic. In an instant, my world had gotten a lot smaller and darker and weirder. It seemed inevitable that in moments I'd be squealing for someone to get me the hell out of there and flailing my limbs (though to little effect, because one can only move so much in an MRI scanner only fifty-five centimeters wide). But just as quickly as the panic came, it vanished upon being lulled into boredom by the groan of the scanner. After two hours, the technologist pulled me out pink-faced and squinty-eyed. I sat on the scanner bed for a moment, trying to reorient myself to real life. *That was awful.*

I told myself that I'd never do another MRI session, but after she handed me $50, I found myself asking how to sign up for more.

I also interviewed to become a tutor at an inner-city elementary school for a work-study program called America Reads. I showed up in my least-wrinkled dress shirt and a pair of slacks that I'd worn only at my ninth grade homecoming dance. It paid $16 an hour, and if I worked the maximum twenty hours a week, I thought I might just make my tuition payments. The interviewer gave me the job right away, but I wouldn't get my first paycheck until the end of February. New sources of income aside, my economic troubles were far from over.

When the first check finally arrived, I decided that the occasion called for a celebratory feast, even if my situation was still precarious.

After my night class, I walked to a Whole Foods supermarket and bought a head of broccoli, two carrots, a clove of garlic, and three red potatoes.

I carried the groceries to my parking lot, made sure no one was looking, and snuck into the van, which was parked underneath a lamppost. The light of the lamppost faintly lit up a section of my van's floor, allowing me to see the vegetables I was about to slice up into my frying pan.

I stuffed the vegetables I didn't plan on eating into one of the plastic compartments of my storage container. (In January and February, the North Carolina temperature was still a cool and consistent 40°F, so I presumed that most of my food would keep without a real refrigerator.) I took out one carrot, half of a potato, and some sprouts from the broccoli, and cut them up into tiny pieces with my knife. I grabbed my water bottle, which I'd filled up on campus, and poured the water into my pot, which I set to boil on my isobutane backpacking stove on top of my storage container. When the water started boiling, I tossed in some spaghetti noodles. And then I mixed in the diced vegetables, a packet of spaghetti seasoning, and a healthy glob of peanut butter, which gave the sauce a thick, almost sweet consistency. I stirred carefully, making sure not to splatter the walls of the van or my dress clothes that hung on hangers on my hook.

The pot was filled almost to the top. It was a steaming, sumptuous feast. I twirled my fork into the thick sauce, lifted the noodles with a few vegetables stuck to the peanut butter to my mouth, and gloried in the tastes. Oh, how I feasted. I feasted until I could feast no more.

In the weeks ahead, I'd tire of spaghetti stew, so some nights I'd eat just cereal. On others, I'd concoct something more sophisticated, like rice and bean burritos. I'd let the red beans soak in the van when I was on campus and then cook them with rice at night, wrapping it all in a tortilla with some tomato and onion.

For breakfast, I'd have either cheap Kroger-brand cereal with powdered milk or a hot bowl of oatmeal (again, with peanut butter). Most of these meals cost me less than $2, and I thought they were healthier, tastier, and more generously portioned than what I could get for $15 at a restaurant. According to my calculations, I paid, on average, $4.34 for food a day.

I couldn't think of an easy way to wash dishes, so I'd scrape as much food from the bottom of the bowl as I could with a piece of bread or a tortilla. I thought of whatever was left in the bowl – each crumb of oatmeal and speck of dried spaghetti sauce – as reminders of my meals' ancestral past, forever seasoning each subsequent dish with its chromosomes like a father passing on genes to his progeny. It was gross, but it wasn't making me sick. And that's all that mattered.

After a week of vandwelling, I could say that I was beginning to figure things out.

I took showers at the gym where I'd also brush my teeth, floss, and shave. At the library, I'd charge my laptop, phone, and camera, and take naps on a couch in the Thomas Reading Room (aka the Chinese Reading Room) on the second floor.

The library would close early on weekends, so I wandered around campus like an ant, sniffing around for anything useful, for any place I could stay, for any extra food I could scrounge. With my Duke ID card, I learned that I could swipe my way into most any building. Now, I could study privately in a warm classroom. Sometimes, when I needed a break, I'd watch streaming TV shows or listen to music on my laptop.

Wherever I went, I kept an eye open for pens or pencils on the ground. I'd reuse old bread and tortilla bags so I didn't have to buy Ziplocs. When I filled up my wastebasket with garbage, I'd simply drop it off in a public trash bin on my walk to campus. At the pace I was using up my clean clothes, I'd only have to go to a Laundromat once a month.

But I had no shortage of discomforts. New, strange, unidentifiable smells greeted me each evening. Upon opening the side doors, a covey of odors would escape from the van like spirits unleashed from a cursed ark. I could never tell exactly what the smell was. It wasn't always bad, but it was never good, and a fourth of the time it was downright foul. This was especially disquieting because I have a particularly weak sense of smell. If I can smell something unpleasant, someone in the next county might be wearing a puzzled expression, wondering, *Someone passed gas here, but I don't know who.* If things got any worse, it was conceivable that the van's smells would absorb into my skin, causing campus administration to assign a team of janitors to follow me wherever I went, pressure-washing the walls behind me.

To deal with the smell, I made my front passenger seat the "laundry area" in order to segregate my rancid workout clothes from infecting my clean, neutral-smelling clothes in the back. I hung my wet towel (after showering at the gym) on the passenger seat so it was in the sun during the day where it would dry.

I bought an air freshener and a broom, twisted off the broom head, and swept out the van.

Sleeping in cold temperatures also took some getting used to. Even though January and February are North Carolina's coldest months, the nighttime low is generally a fairly reasonable 30–35°F. One night, though, it got down to 10°F. I didn't have a heavy coat, so I delayed leaving the library as long as I could. I finally got to the van at 3 A.M. It was about as cold inside as it was outside, though the van did offer protection from the wind. I stripped off my school clothes so I could shiver into my sleepwear. Nearly naked, my knees clanged against each other, my muscles tensed up, and my penis turtled into my body. I put on my hat, gloves, an extra pair of wool socks, thermal underwear, and my pajamas. With chattering teeth and fingers as hard and stiff as icicles, I struggled to get a grip on my sleeping bag zipper. Within moments, though – now burritoed in my own body heat – a warmth radiated across my arms and legs and torso, as if my bag were stuffed with sunrays. The cold on my face and the warmth of my sleeping bag make for prime sleeping conditions – it was a natural tranquilizer that, within moments, turned my eyes heavy and caused me to drift away into dreamland.

Oh, how I loved sleeping in the cold. I'll deal with frostbitten black bananas, frozen jugs of water, and the windshield coated in icy exhalations in the morning, so long as I get to revel in this warmth in the midst of cold. Sometimes I wished for the temperature to be warmer, but why live in a chronic state of want, constantly hoping for heat in winter and cold in summer?

The next morning, though, I wouldn't have minded a heater of some sort. It still felt like it was 10°F in the van, and the prospect of getting out of my sleeping bag and thermals and into my school clothes was a duty that required a level of willpower that I found difficult to rouse. I was going to be late for work, so I feverishly stripped off my pajamas, shivered into my clothes, and lit up the stove for my morning oatmeal. While I was daunted by the prospect of being exposed to the cold in this manner every morning, there was no question about it: I'd take this over debt any day of the week.

While I was dealing with the cold and smells and hunger well enough, and while I had a job with a steady paycheck, I was still always one accident, one injury, one overlooked expense from going bankrupt. I thought constantly about my goal.

—Day Fifty of Vandwelling Experiment—

18

....

MY FIRST GUEST

SAVINGS: $1,160

WHAT WOULD HAPPEN IF they caught me? Would they pity me, laugh it off, and let me go? Or would they get all bureaucratic and corporate on me and say that living in a van was against the rules?

It wasn't just my experiment that was at stake. My freedom, independence, and comfort were, too. If I got caught, I'd have two options: I could find some other radically cheap dwelling, or I could drop out of school. What wasn't an option was breaking the promise I made to myself months before: I was not going to go back into debt. No matter what.

I knew from my Brooks Range mountain climbs that to get to the top of a mountain, you have to be half-insane. The climber must approach his goal with a zealotry that may be inappropriate for normal, mundane things but is essential for the grandiose. If I was going to stay out of debt, I knew that I would need to take a similar approach.

So every time I saw the light of a car's headlight splash into the van at night, or heard someone – perhaps a security guard –

walk around my van, I was terrified – terrified not just of getting caught, but of having to actually "walk the walk" and do what I promised myself I'd do. I knew that, if it came down to it, I'd live in the woods or get my food from Dumpsters, which, though frugal styles of living, were things that I by no means wished to experiment with, as I'd begun to embrace the many comforts and conveniences the van offered. I certainly didn't want to suffer needlessly in a less-pleasant dwelling if I didn't have to. I felt paranoid constantly. I needed the van. And I needed to keep it a secret.

The Mill lot, unbelievably, turned out to be well suited for my experiment. While it was next to a busy street, it was also really far from campus, making it one of the least visited lots at Duke, offering me some precious privacy.

Still, though, I had to be careful. When I walked to the van at night under the bright streetlamps of Ninth Street, I felt like I was starring in a film noir. If I saw anyone in my parking lot – either in their car or on foot – I'd walk past the van and wouldn't return until I was sure nobody was around. Sometimes I'd stand in the dark with my back against the shadowed brick walls of the adjacent apartment complex while waiting for students to walk past me so I could scurry to the van when the coast was clear. Or I'd sit on the porch steps of the apartments pretending to read a book, surveying the scene from underneath my jacket hood, eyeing anyone and everyone who might catch sight of me. And when the students had gone inside their apartments or to the bar, I'd casually saunter up to the van, quickly unlock the front passenger door, reach inside to gingerly unlock the side door, and gracefully hop into the van and slam the door shut in one practiced, elegant motion. Once inside, I moved furtively, placing my steps in spots where I knew the floor wouldn't squeak. In the morning, I was just as careful, lifting the blinds before exiting to make sure no one would see me leave. I did this every morning and night.

Early on in my experiment, I knew I had to develop a strategy to keep the van a secret. Hence, the seven rules of vandwelling:

1. Do not talk about vandwelling.
2. Do not talk about vandwelling.
3. Never leave the ignition running for light or heat or any purpose.
4. Don't make any unnecessary noises inside.
5. Close all windows when cooking so no one hears the patter of pots or the tinkling of silverware.
6. Never, ever let anyone see you enter the van.
7. And never, ever let anyone see you leave it.

For the first two months, I hadn't told a single soul in Durham about the van—not because I didn't want to (I was *dying* to tell someone), but because I couldn't. Or at least I thought I couldn't.

I wondered if I was being a little *too* obsessive. Maybe a little *too* secretive. Maybe fellow students wouldn't care about sharing a parking lot with a vandweller. Maybe Duke administration wouldn't care. Maybe they'd even support one of their students exploring a new way of living. But my every instinct told me my experiment would not be so well received.

"You're going to get in so much trouble when they find out, you know!" my mother would remind me, adding, "*It just isn't normal!*"

At first, I was fine with the idea of social isolation. I figured the van would give me a chance to write, think, read, reflect, and grow, just like Thoreau's woodland cabin did for him. My solitude—I hoped—would trigger a personal renaissance. But after two months of being around people and not connecting with any of them, I became surfeit with solitude. I hadn't had a single real, genuine conversation with anybody. And whenever I got close to having one, the topic of where I lived would come up and I'd have to sour the conversation with preposterous lies.

Soon, it just became easier to avoid conversations altogether.

It was fitting, I thought, how my first want wasn't a product or a luxury; it was companionship. I was doing without just fine. I didn't need stuff, I didn't need fancy foods. I'd given up TV and beer and going out. But, oh, what I would have given to be drinking with my high school buddies back in New York. Oh, what I would have given to talk with just one of the pretty girls in the library. I thought of Sami constantly.

But my solitude was self-inflicted. I was the one who chose to live in a van. I was the one who chose to end our relationship so I could go back to school. I was the one who was selfish.

What was this terrible urgency to always be alone? Most of my day was spent sitting in the corner of the library on my laptop. You broke it off for *this*? Why did I have to always be going somewhere else, leaving someone behind? I always heard "the road" calling me, but why did I listen? The road was always so lonely and miserable.

Yet I couldn't stay with Sami, or any girl for that matter. Maybe it was because I'd once felt restrained by debt, so now I'd get antsy whenever I felt restrained by anything else. I'd become obsessed with freedom. I could sense the slightest abridgment of my freedom like a princess who can feel the impression of a pea under forty featherbeds. I felt it when I was in romantic relationships. I felt it when I was given a gift. I felt it when someone held even the faintest influence over me. And when I felt it, I felt rage – a heart-thumping, roiling rage in the pit of my chest that was so overpowering I had to talk myself out of rashly fleeing and separating myself from that which was infringing on my freedom. When is it going to end? When can I be a normal human being?

It was a curse, this always needing to be alone. I used to think the urge would go away after some big trip or adventure, and that a mountain climb or hitchhike might somehow scratch my itches, calm my nerves, lull my wanderlust, and grant me, finally, a peace of mind that would permit me to settle down and be content. But this was never the case. I was like a soldier who, upon completing his tour of duty, wanted nothing more than to go back to the front lines.

In the library, I popped a DVD into my laptop that Sami had made for me when we were together. It was a series of pictures of us together in Mississippi, on our hitchhike, and in Alaska with Jack Johnson's song "Better Together" playing in the background. (*Yeah, it's always better when we're together . . . Yeah, we'll look at the stars when we're together . . .*) Having nowhere else to watch it, I turned it on in the corner of the library, wearing a pair of the library's borrowed headphones. As I watched images of us holding hands in Vermont, I had to hide my face behind the screen to make sure that no one saw the tears coursing down my cheeks.

As much as I may have desired a companion—particularly a romantic companion—I knew I had to stay away from girls. I *could not* fall in love again. That would have defeated the whole purpose of going back to school. Yet my decision was beginning to take its toll. By the eighth week, I began to sing and talk to myself with unprecedented frequency. By the tenth, I began to actually *converse* with myself. One such conversation:

Me: It smells in here. You should clean up after yourself.
Myself: But, baby . . .
Me: Don't you "but, baby" me!

Because it was only a matter of time before I'd buy a volleyball and paint a face on it with the blood of my palm, I knew I had to start reaching out to other people. I had to tell *someone* my secret.

"Hey, man, you from Durham?" he asked. He had scraggly blond hair and a feral beard, and was wearing enough denim to exceed what I thought was fashionably appropriate.

Normally, I recoil from contact with the homeless with the same sense of urgency that I hide from a pair of well-attired and determined door-to-door Jehovah's Witnesses, but I greeted this guy — who was in front of the Whole Foods grocery store on my walk to school — as if he were a long-lost high school acquaintance.

"Uh, no, I'm not really from around here," I said. "But what's up?"

He told me he was new in town, couldn't find work, and was on his way out. He assured me that he wasn't a "wino like the rest of them" and that he just needed some money for food. I didn't know what to believe, but I felt sympathetic, and I thought this was my chance to finally have a real conversation with someone who wouldn't rat me out.

"It's hard in this economy, y'know?" he said. "A guy can't get a job. So . . . would you mind lending me a couple bucks?"

"Yeah, I know how it is, man," I said, trying to establish a common bond. "In fact . . ." — I paused with deliberate graveness — "I'm living out of my van."

I hoped that we'd then trade declarations of sympathy and swap tales of poverty, but I was surprised with how indifferently he passed over my admission and continued to prod me for money.

As much as I prided myself for living "on the edge," I had to admit that I wasn't homeless in the way that he was homeless. The voluntary nature of my experiment prevented me from experiencing poverty in its most extreme and authentic form. I wasn't poor, of course. Real poverty is having no way out. Real poverty is what one-fifth of the world lives in. No matter what, I had food to eat and a warm place to sleep every night. If I needed to, I could go back home to my parents, stay with close friends, or get a job in Coldfoot again. I had an education, connections, a good upbringing. Real poverty has little to do with being broke. Real poverty is not being able to change your circumstances. I was playing with poverty; he was living it.

I gave him $3. He thanked me and said he was leaving town to find work.

The next day, I heard the same voice yelling in my direction.

"Hey, man, you from Durham?!" he said, now with a slur.

I'd been had. This was the prologue to his standard routine, I realized. I ignored him this time, but I couldn't help but notice that he was warming his hands around a flaming barrel, huddled together with his council of homeless comrades.

I had no sympathy this time around, just disdain, maybe even a strange hint of envy. Whenever I'd see him again, he'd ask me for money, and I'd want to throw my fist into his jaw and kick him when he was down, sending his teeth scattering across the asphalt.

I felt like I was some leprous pariah, some low-caste untouchable. While I identified with Thoreau on many levels, his stance on sex was more than foreign to me, and he was far from a source of guidance. Thoreau, who had the libido of a turnip, was assumed to have died a virgin. I, however, had to contend with the insatiable desires of a raging twenty-five-year-old male's "I will do anything that moves" libido. I thought of women constantly. Being around so many pretty ones, and feeling that I had no chance with them, felt like some cruel Dantean punishment, as if I was paying for the sins of a tragic life of lust, destined as I was to be eternally tormented by their charms. Feeling incapable of wooing one of these girls made me crave them all the more; it was as if I needed a woman to like me to validate my presence on this earth as a living, breathing, existing human being.

I couldn't tell anybody about the van, but would it make any difference if I could? Why bother trying to meet somebody? What girl here would want to be with a guy who lives in his *vehicle*? While such a home might seem novel, or cool, or just plain ordinary in Alaska or at a hippie college out west, living in a van at a place like Duke was just plain *weird*. Confessing such a thing would be like revealing that my privates were pocked with some hideously contagious and suppurating STD.

But it wasn't just the van holding me back. While I thought I'd experienced enough of the world to no longer be vulnerable to inferiority complexes, I couldn't help but be intimidated by how all these young people were just so *accomplished*. Many spent their adolescences getting refined Victorianesque educations at private schools: playing instruments, excelling at various artistic media, and spending a suspiciously large amount of their hours volunteering. (I met one freshman who had published a book of poetry when she was nine!) As an adolescent, apart from three weeks of trombone lessons in the fourth grade, I didn't do anything to "cultivate" myself. While Duke students might have spent their childhood summers touring Europe or mastering the art of jumping at equestrian camp (I'm exaggerating, but go along with me), as a boy I spent mine watching the USA channel's weekday afternoon lineup, highlighted by reruns of *Quantum Leap* and *American Gladiators*.

Everyone was just so wealthy, too. At Duke, the average white student's family income was $230,000, and those of other ethnic backgrounds weren't far behind. My parents had blue-collar jobs and modest salaries for which I felt no shame, but one can't help but feel out of place when many students have brand-new SUVs of their own.

I was dying for human contact. I was willing to share my secret with someone, but everyone I met at Duke just seemed so radically unlike me. While I still hadn't had any real conversations, I was able to – through a few superficial exchanges and some borderline stalker-ish eavesdropping – get to know a few of my fellow students.

Alicia was an ambitious, overachieving undergraduate political science major and a president of three clubs who set multiple alarms on her watch so she'd know when to commence doing homework for different courses. Per her father's expectations, she was applying to thirty-five law schools – yes, thirty-five! – for the next fall semester. Despite her law school ambitions, I heard her confess to another classmate that her real dream was to kayak the Amazon and hike the Pacific Crest Trail. When I happened upon her at a coffee shop, I went out of my way to bring up her dream trips and how "awesome" they sounded. She seemed happy to receive support, but she mentioned, regretfully, that she was also trying to get into law school. It seemed like an easy call for me. The law school path would surely pile on top of her shoulders a decade's worth of debt that would prohibit her from going on the long journeys she wanted to go on. The adventurous path, on the other hand, could allow her to embrace a freedom she may never otherwise get to acquaint herself with.

Kim was a thin, toned, gym-obsessed Ph.D. grad in the neurobiology department. Her program demanded a five-year commitment, summers and most holidays included. While she received free tuition and a generous stipend, she also had to work countless hours a week on top of her rigorous study schedule. When I asked why she chose neurobiology, she told me she had "no idea," adding, "Sometimes I don't even feel like my life is mine." She said she was good at neurobiology as an undergrad, so she chose to apply to a Ph.D. program for lack of any better ideas. I was taken aback by how indifferently she decided to enroll in a Ph.D. program. How could someone make such a life-altering decision with so little conviction and forethought? I met students everywhere who had similarly devoted half their twenties to something they had little passion for. Eventually, they'd be hyperspecialized and would only be qualified for work that didn't at all fit their true interests or character.

While Sami had no college education, it was clear to me now that, in ways, she was smarter than a lot of these Duke students. At least she knew to follow her passions and live the precious now instead of preparing, preparing, preparing – constantly preparing – for something that may actually turn out to be not that great tomorrow.

I met Scott, a pot-smoking, Adderall-popping undergrad at a coffee shop. Scott loved rock climbing, and when I told him that I'd gone on a really long hitchhike, he was clearly envious and hungry to go on such an adventure of his own. Yet he'd already landed a prestigious internship with Morgan Stanley (or one of those companies in the financial services industry), which he wasn't at all enthusiastic about. I asked him why he was about to take a job he didn't want, and he responded, "I got to live, right?"

While my life was far from perfect, the van was making it quite clear to me that a human being can live on very little. I spent half my day in a home where I had no air-conditioning, no heat, no plumbing, no electricity, no magazine subscriptions, and no Netflix account, yet I was still breathing and standing and living just fine.

Most disconcerting of all were the whispers I'd constantly hear of Citibank, Goldman Sachs, and Wall Street – where many Duke students go to work after graduation. Duke takes pains to sell itself as a traditional liberal arts school, but like a lot of schools nowadays, it has strayed far from its liberal arts roots. While they don't offer an undergraduate business major like most schools, economics, a close cousin to business, is the most popular major year after year.

Without any better ideas, I thought that I might have some luck finding friends in the student outdoors club. I spent money on my first nonnecessity: a $50 membership fee, which would grant me access to their climbing wall in the gym, as well as a few subsidized trips to the woods and mountains of North Carolina.

The club was heavily populated with grad students from the Nicholas School of the Environment, a department that, I figured, would be a surefire hotbed of liberal-minded do-gooders. Together, I imagined us practicing eco-terrorism in the late hours of the night, perhaps sawing down billboards and blowing up dams. On weekends we'd attend wild parties populated with dread-headed, tie-dyed classmates who loved scrubbing oil off sea turtles and fornicating in body-painted orgies. But instead, I quickly learned that about half of the environmental majors were getting their degrees in order to obtain jobs in the oil and gas industry.

Goldman Sachs? *The oil and gas industry?!* What the hell was going on? What's the point of schools like Duke if they're

merely funneling grads into careers that—excuse the colloqui-
alism—fuck shit up?

Whenever I came across a student who was headed to Wall
Street, I wanted to ask: "You actually want to work for one of
those companies?! Aren't they kinda evil? *Wall Street?* You mes-
sin' with me?"

They weren't, though. Every May, it's as if Goldman Sachs
and Bank of America and companies like them position a giant
vacuum at the end of Duke's and other elite schools' commence-
ment stages, sucking up many of its cap-and-gowned grads to
be emptied out on Wall Street.

Reading sixteenth-century French poetry, suffering through
Kant, and studying the finer points of the Jay Treaty may seem
to be, on first appearance, completely, utterly, irrefutably pointless, yet somehow in studying, discussing, and writing about
these "pointless" subjects, the liberal arts have the capacity to
turn on a certain part of the brain that would otherwise remain
shut off—the part of our brain that makes us ask ourselves questions like: Who am I? What's worth fighting for? Who's lying to
us? What's my purpose? What's the point of it all? Perhaps many
students would rather not be irritated with these questions, yet
being compelled to grapple with them, it seems, can make us far
less likely to be among those who'll conform, remain complacent, or seek jobs with morally ambiguous employers.

Some students who enroll in business and vocational programs don't get the chance to ponder such questions. After four
years of creating models and punching numbers, they often
leave college with a startling unwillingness (or inability) to ask
themselves necessary ethical questions. And when students go
to school for the sole purpose of getting careers and making
fortunes, the degrees they leave with may no longer be flimsy
rolls of parchment but dangerous weapons. More and more, I
started to believe what Desmond Bagley said: "If a man is a fool,
you don't train him out of being a fool by sending him to university. You merely turn him into a trained fool, ten times more
dangerous."

When I was applying for admission to Duke, I had a ludicrous vision of what my education would be like. I figured Duke would be where students questioned everything, dreamed up radical and revolutionary ideas, and reevaluated our economic and social institutions. And while I was getting some of that from my liberal studies courses, I learned that the consumer-capitalist model not only goes unchallenged in most university curricula but that it's quite literally taught. The university today is not a place where we go to question the dominant institutions; it is a place where we learn to support them.

I didn't know why all this bothered me so much. Why did I care how other people lived their lives? Did it really matter whether they went into debt or worked for Goldman Sachs? What did any of this have to do with me? Maybe I shouldn't have cared, but I couldn't stop myself from caring. Maybe it was just that I didn't like to see lives squandered the same way I hated to see good food thrown away. Maybe I thought they'd be happier if they got off the fast track to careers and out onto the meandering river to nowhere in particular. Maybe I felt this would be a better, smarter, more democratic world if we had more poetry-loving citizens than money-hungry careerists. Or maybe I was just lonely.

I decided I needed to connect with someone. Someone who cared nothing for the career-obsessed future, only the passionate warm-bodied present. I figured I ought to find some middle-aged divorcée with whom I could have a fling – someone who'd just emerged from a rocky, abusive relationship; someone off campus who couldn't care less about Duke and debt and careers. In a moment of desperation, I scanned the "women seeking men" ads on Craigslist for the first time in my life.

I could just picture it: Things would get off to a promising start. We'd meet up at a coffee shop, and she'd be delighted to have found a symmetric, disease-free student – a passably attractive young man who wanted nothing more than to lavish her aged, though alluring, body with compliments and caresses. (*Finally a man who's got his shit together,* she'd think.)

The romance in the air, however, would turn fetid when I'd have to sheepishly present my "home" to her.

"Honey, you've got to be kidding me," she'd say, shaking her head. "This is just my luck."

"Hold on, hold on," I'd say. "Have you ever heard of a guy named Thoreau?"

"Lord," she'd say, scolding the sky, "if there's an able-bodied man with a job left on earth, you're sure doing a good job hiding him from me."

It was a bad idea. Ad titles like "I need a man who can satisfy me" were asking for more than I could offer anyway.

I spent every moment of every day alone. I played basketball alone. I watched movies alone. I sat in the library alone. If I could have afforded it, I'd be the guy in the cafeteria eating alone. When the library closed, I'd go to a classroom in a quiet building that was open 24/7. There, I'd blare loud music on my computer and dance to Old Crow Medicine Show's "Wagon Wheel" alone.

And then, abruptly, I had a guest in the van.

I heard a tinkling of silverware, a crinkling of plastic bags. I knew exactly what this meant.

There is a mouse in my van.

For the next eight hours, I could do nothing but tightly seal all orifices and eye the ceiling where I heard it (or them) running to and fro. In the morning, after only a few hours of sleep, I, exhausted, noticed little black pellets—a third the size of Tic Tacs—scattered everywhere: across my floor, atop my storage container, and—in more concentrated quantities—inside my waste basket and unwashed pots.

I should have noticed the pellets the previous day, but I had probably dismissed them as remnants of other foods I'd eaten, like the meteor showers of rye bread and cracker crumbs that had sprinkled from my lips to the floor.

I tried to forget about it. I'm used to squalor, I reminded myself. Why should I let a little rodent ruin my night?

Yet I couldn't stop myself from worrying about it. I imagined one of the bolder members of the colony slinking into my sleeping bag at night to tour my body like an explorer charting an exotic peninsula. Perhaps I'd come back to the van the next day to see that, upon making use of my food, they'd multiplied tenfold. Swarms, several layers thick, would be writhing libidinously in a gelatinous orgy. There'd be unrestrained fornication, mothers would eat their offspring, and several of the more acrobatically inclined would twirl on coat hooks and scamper up my hanging partition.

On the second night of the invasion, as I was just about to fall asleep, it hit me. I had cereal the previous morning. Were those cracker seeds at the bottom of my cereal bowl or were they . . . ? Oh no.

I think I ate mouse shit.

The next day, I threw out food the mouse had been in, bought several mousetraps, and slathered them in peanut butter.

But it didn't take the bait. I could hear it scurrying in my walls and ceiling. I lay there listening, nervously darting my eyes toward each rustle, scrape, and squeak. I could see the impression of its feet in my ceiling upholstery. It was everywhere. Running back and forth, front to back, side to side. What the hell was it doing? What grand schemes did it have in mind? Its diligence was terrifying.

At one point, in a fit of anger, I stomped my boot against the ceiling from end to end in hopes of tormenting it out of the van. Then I got out my frying pan and did the same.

This time it was beneath me. I saw it on my floor. It turned into a dart of fur, flying up into a crack in the ceiling upholstery. I had it cornered. The ceiling upholstery fluttered desperately. I whacked away with my frying pan until it stopped.

I was losing sleep from the mouse and an overabundance of schoolwork. I was doing MRI studies multiple times a week and tutoring at the elementary school almost every weekday. I felt like I was beginning to revert to the bad habits of my undergrad years. I stopped working out at the gym, and I even spent a couple of dollars on coffee to help me pull all-nighters.

In the library, my head started throbbing. Menopausal hot and cold flashes were making the hairs on my arm rise and fall like time-lapse footage of flowers responding to the sun. I became dizzy and delusional.

I zigzagged back to the van and pulled the sleeping bag over my sweaty brow. What had done this? My unwashed fork? The month-old bottle of squirtable butter? The mouse turds?

I positioned my wastebasket next to my bed and christened it with a few introductory heaves—a mere preamble to the story that would follow. My throat, like a fire hydrant uncorked by a group of overheated inner-city juveniles, discharged the entirety of my stomach's contents in one impressive burst. Unsure if I was feeling better, I plopped my sweat-soaked head on my bed. A heavy rain began to fall. Near the back of my van, where the water would pool on the roof, I looked up at a small nickel of wetness in the ceiling upholstery that slowly grew to a pancake's diameter. Then the rainwater began to pitter-patter on the back of my thigh.

What am I doing? I thought.

I was sick and lonely. There was a hole and quite possibly a dead mouse in my ceiling. I wanted the van to show me how to live. But it seemed like I was missing several essential ingredients. And living in a van wasn't some sort of model of living that I could use to show people how to live more sustainably. While I didn't go out much, I was still driving a vehicle that guzzled gas. I shopped, not at farmers' markets, but at large, inexpensive supermarkets that paid their workers poorly. I outfitted the van with products from Walmart.

I found that I couldn't do without people the same way I could do without luxuries and material comforts. I thought of Thoreau and how he called the "neighborhood of man" around him "insignificant."

Bullshit, I thought. He said that he communed with pine needles and dewdrops, hardly mentioning that he constantly had visitors, or that his family and friends were just a short walk away. I thought that if he knew real solitude – not knowing anybody, not being able to talk to anybody, and not having anyone to relate to – then he wouldn't have written such nonsense.

—Day 115 of Vandwelling Experiment—

19

....

SOLITUDE

SAVINGS: $1,771

AFTER VOMITING WHAT FELT like half my body weight, I laid my clammy head on the bag of winter clothes that I used for a pillow.

Even though no one was there to see me, I felt embarrassed. It was a pitiful scene. I was lying in a van half-naked. Next to the bed was my wastebasket, full of creamy sewer-green vomit. I had no one to help me. The nearest bathroom was a quarter mile away. There might be a mouse carcass in my ceiling.

I couldn't help but bring my whole experiment into question.

I woke up groggy the next morning but in far better shape. It was a weekend, so my parking lot was empty, giving me the privacy I needed to dump my wastebasket's contents under an arborvitae tree behind the van. When I stepped back inside the van, I got a whiff of something sour, sweaty, and moist, reminding me of a sopping-wet bag of hockey equipment.

I palmed my ceiling like a mime in a box, worried that I'd come across a bump, which I sure enough did. I donned a pair of gloves, lifted the upholstery, pulled out the flattened mouse carcass, and threw it under the same tree.

I cleaned out the rest of the van from top to bottom. I swept out the crumbs and mouse turds. I scrubbed the dried tomato paste off the top of my storage container, and I put my pot, pan, and cereal bowl into a bag that I'd take to campus to wash in one of the bathrooms. I shoved all my dirty clothes and bedsheets into my foldable laundry hamper, which I hauled to the Laundromat across the street. It was too hot to stay in the van, so I brought a tarp to East Campus and lay on it under an oak tree, hoping the sunshine and fresh air might hasten my recovery. Later, I went to the gym, where I played basketball, sat in the sauna, and scrubbed myself clean in the shower before shaving my face and brushing my teeth.

I was determined to recuperate, not just for my own well-being, but also because I needed to get myself into presentable shape for a five-day field trip to a biological station in Highlands, North Carolina, for my Biodiversity in North Carolina course. The day before I was to go, my mother sent me an e-mail reminding me about my tax return. My tax return? *My tax return!* I'd completely forgotten that I was getting a tax refund!

It was a $1,600 golden ticket that Uncle Sam was going to slip into my bank account.

I was rich.

I knew then and there that I was going to get through the semester debt-free.

At first, I felt relief. But then there was ambivalence. This was a turning point, I realized. I had financial security for the first time in months, and I knew I might be financially secure for good, because in just a few weeks, I was heading back to Alaska to work at the Gates of the Arctic for another season.

I'm not poor anymore, I thought nostalgically.

At the biological station, I shared a room with Chuck, a forty-six-year-old student and former accountant who'd quit his job so he could hike the Appalachian Trail and enroll in Duke's liberal studies program. In our room, we swapped hiking stories, mused about Thoreau, and described our final papers, both of which were about the importance of wilderness.

For the course, we all went on daily field trips to see different biozones in the Appalachians; listened to lectures given by conservationists, salamander experts, and biologists; and learned how to identify all sorts of trees, mosses, and fungi with our magnifying glasses. In the lab, under microscopes, we looked at the tiny, crawling forgotten kingdoms I never knew existed.

Knowing that my tax refund was coming, I slackened some of my strict spartan standards: I bought a case of beer for me, Chuck, and a couple of other classmates; I mailed Marietta, the woman who let me stay at her house in Durham, a $50 gift certificate to a fancy restaurant; I dined at a restaurant twice; and I slept in a heated room on a comfortable bed.

These were the sort of purchases and pleasures I commonly indulged in years ago, but treating myself to them now made me feel a strange sense of guilt, as if I'd cut some corner I promised myself I wouldn't cut. During my third night at the station —beleaguered with self-reproach—I dragged my sleeping bag outside and slept on the pavement under the stars.

I didn't need these things, I thought. The beer, the food, the bed. I didn't even really want them. I was buying stuff, not out of need, but simply because I could afford them.

Before I enrolled at Duke, I decided not to take out loans because I knew that if I allowed myself access to easy money, then I'd again fall victim to the consumerist trap. I'd be indiscreet with my money. I'd begin to pay for and rely on things I thought I needed but didn't. I'd lose perspective.

If we put a man in a country club, he'll suddenly feel the need for a yacht. But if we put him on a solitary island, his only desires will be those essential to his survival. I wanted to keep my needs simple. I didn't want to lose perspective. I didn't want to once again be swallowed whole by the dominant culture, accepting its norms and values and desires as my own.

I knew what I was missing in my life. It wasn't things. It wasn't heat, plumbing, or air-conditioning. It wasn't extra space, or an iPhone, or a plasma screen TV. It was people. It was a community. It was a meaningful role to play in my society.

On our last night at the science base, we had a bonfire in the woods. In the early hours of the morning, it was just me, Chuck, and two other male students: Joe, a thirty-one-year-old yogi, and Salman, a Pakistani with an interest in theology.

After a few beers, the mood turned jovial and the conversation meandered. Eventually, the inevitable question arose when Chuck asked, "Ken, where are you living?"

Normally, I would have said, "On Ninth Street," and that would have been the end of the conversation. This time, though, I said, "I'm living in my van. I've been living in it this whole semester. I'm trying to stay out of debt."

"What does he mean?" asked Salman.

"Ken's just said that he's been living in his van," Joe clarified.

"It's not so bad. I wash at the gym, get electricity in the library," I said.

"Does anyone know at Duke?" Joe asked.

"Not that I know. God, I hope not."

And that was that. My secret was out. Salman had an expression on his face that one would normally reserve for an alien encounter. Chuck and Joe seemed only mildly amused.

"How do you iron your clothes?" Salman asked.

"Ha. I don't, I guess."

I went back to Duke and I felt summer's presence. A half dozen dogwood trees that shadowed my parking lot bore branches heavy with thick, lustrous white flowers. They buzzed with a million bumblebees and smelled of a woman's hair. On mornings, I awoke to a medley of birdsong so loud and cheery you'd think my little hermitage was tucked away in a copse of trees at Walden Pond. At night, I was whirred to sleep by the chorus of cicadas. During rainstorms, I listened to millions of raindrops drum against the roof and watched them wiggle like sperm down my windows. My van began to feel less like a novelty and an experiment and more like a real home.

And now that I'd been through almost a full semester in the van, I couldn't help but observe improvements in my physical condition. Because I'd almost entirely cut out meat, dairy, and beer from my diet, and because I visited the gym frequently, I was leaner than ever. After my mouse and throw-up incidents, I kept the van clean and tidy, and washed my pans more frequently. I never got sick again, and something about sleeping with the climate rather than against it gave me a hardy constitution. I handled the cold with nonchalance. I dealt with the heat with gentlemanly sportsmanship, laughing with it good-naturedly when it got the best of me. I didn't care as much about being hungry, or being deprived of domestic amenities, or even when insects roamed my body when I lay beneath the giant willow oaks on campus. Discomforts are only discomforting when they're an unexpected inconvenience, an unusual annoyance, an unplanned-for irritant. Discomforts are only discomforting when we aren't used to them. But when we deal with the same discomforts every day, they become expected and part of the routine, and we are no longer afflicted with them the way we were. We forget to think about them like the daily disturbances of going to the bathroom, or brushing our teeth, or listening to noisy street traffic. Give your body the chance to harden, your blood to thicken, and your skin to toughen, and you'll find that the human body carries with it a weightless wardrobe. When we're hardy in mind and body, we can select from an array of outfits to comfortably bear most any climate.

Yet what physical changes I saw take place throughout my experiment paled in comparison to the financial revolution that toppled old preconceptions of saving, borrowing, poverty, and wealth. In the van, I'd saved hundreds, no, thousands of dollars. Four months' rent for a modestly priced apartment, plus utilities, would have cost me more than $2,000 for the spring semester. Over the past several months, I hadn't paid any rent or bought any clothes, appliances, or material items. I ate for $4.34 a day, and I lived on $103 a week, not including tuition and school fees.

Normal Living (per month)		Van Living (per month)	
Apartment	$1,000	Apartment	$0
Utilities	$50	Utilities	$0
Vehicle costs/repairs	$450	Vehicle costs/repairs	$71
Gas and motor oil	$150	Gas and motor oil	$23
Entertainment (TV, Internet, video games, travels, clubs)	$75	Entertainment (TV, Internet, video games, travels, clubs)	$33
Miscellaneous	$75	Miscellaneous	$65
Food	$200	Food	$132
Car insurance	$125	Car insurance	$46
Cell phone	$70	Cell phone	$37
Clothes/household appliances	$100	Clothes/household appliances	$0
Total	$2,295 ($27,540/ year)	Total	$407 ($4,884/ year)

My meal plan—let's call it the "spaghetti stew meal plan"—would have cost me $1,644 over the course of an entire year. Yet in 2012, one of Duke's dining hall meal plans would cost some students as much as $5,780 for just a six-and-a-half-month academic year, or more than $25 a day.

Room and board for an average student at an average school costs a staggering $8,500 for an eight-month academic year. I'd learned from my own experiences that living doesn't have to cost so much. For just the bare essentials (food, vehicle, and miscellaneous costs), I determined that it would cost me less than $5,000 to live for one whole year in a van.

Minimum Living Totals by Type of Dwelling[1]					
Dwelling type	Rent or mortgage/ month	Food/ month	All car costs/ month	Misc./ month	Total/ month
House	$1,000	$137	$0	$65	$1,203 ($14,436/ year)
Dorm	$582	$482	$0	$65	$1,129 ($13,548/ year)
Apartment	$550	$132	$0	$65	$752 ($9,024/ year)
Van	$0	$132	$200	$65	$402 ($4,824/ year)
Tipi/tent/ other	$0	$137	$0	$65	$202 ($2,424/ year)

I'd survived the semester just fine without heat or air-conditioning or restaurant food. I didn't melt into a useless puddle of goo for not having the latest technological doodad. The van had

---

1 It may be unrealistic to not include cell phone, transportation, and health insurance, as well as other costs, but the table is really only meant to illustrate the extreme difference between the amount of money required to live in various kinds of dwellings and also how little we have to spend to live if we cut out the superfluous.

reinforced what I'd already come to know and felt from the bottom of my heart: We need so little to be happy. Happiness does not come from things. Happiness comes from living a full and exciting life.

After spending my morning tutoring at the elementary school, I'd drive home to the Mill lot, hop in the back, sling up my black partition, and sink into a deep slumber. I'd wake up from my nap content, knowing the rest of the day was mine.

I was a monk, and the van, my monastery. (It's amazing how a little financial security can put one at ease.) I would lie in bed for hours, reading, thinking, doing nothing, basking in solitude for as long as I wished, staring at the ceiling, idly musing, unworried about feeling industrious or useful. I pondered everything from the Milky Way to the fallen crumb on my floor. After a run, sometimes I'd sit in front of the gym for an hour or more crumpling dried leaves in my hand. Some kind passerby would occasionally ask, "Is everything okay?" as if my being alone and lost in thought were symptoms of depression. "I'm good," I'd say. "Thanks."

I missed Sami, but it was clear that our decision to separate was the right one. She was doing well at a community college in San Francisco, and I was beginning to reap the rewards of focused study at Duke. I still wanted companionship, but I knew I could wait. The isolation and focused study and idle thinking were paying off. I knew I'd have to change my reclusive policy someday, but for now, I was embracing "solitude as a bride," as Emerson suggested, because it was she and only she who'd help me grow in this season of my life.

Loneliness, no doubt, caused anguish, but over time, solitude gave me something I never would have expected: a culture of my own.

There was no envy in this culture. In years past, I was as susceptible to envy as anyone else: always wanting someone else's car, someone else's friends, someone else's life ("If only I had what he had . . ."). Envy is a bitter fruit, but one that only grows when we water it with the nourishment of society. Remove society, and it will wither on the vine.

I was no longer influenced by cultural fads or other people's values. While hardly anyone at Duke knew about the van, I didn't care what people would have thought about my living in it. I stopped feeling anxious about wearing faded shirts or worn jeans that I'd bought from the Salvation Army. I stopped caring if my hair was too long and out of style. I'd keep my appearance fairly trim and conventional – and the van a secret – but only so I could keep my job and parking lot. I wouldn't feel a sense of shame if some representative of society disapproved of my looks or lifestyle. Rather, I'd wear my poverty as a badge of honor, as a symbol of my unwillingness to dance to the direction of some corporate puppeteer.

Why should I listen to society? Society – as far as I was concerned – was insane. To me, society was boob jobs and sweaters on dogs and environmental devastation of incalculable proportions. We do not listen to the lunatic on the city corner who screeches every day about how the world is going to end, so why should I stop and let society shout nonsense in my ears?

These are society's definitions of poverty and wealth: To be poor is to have less and to be rich is to have more. Under these definitions, we are always poor, always covetous, always dissatisfied, no matter the size of our salary, or how comfortable we are, or if our needs are in fact fulfilled.

When I saw people in their flashy cars or expensive clothes, it wasn't envy I felt anymore. I felt sorry for them. They were obsessed with what other people thought, swayed into accepting the latest fashion trends; deluded by advertisers, marketers, and profiteers; and corralled and branded and shorn of their money. This all revealed an unbecoming manipulability, a lack of real, hard character.

The van kept me debt-free. It kept me dry in rain and warm in cold. It kept me free. How could I feel embarrassed about it? Why should I care what other people thought?

• • •

After our trip to the biological station, Chuck and I began hanging out. We went to a free symphony at Duke Chapel, I visited his apartment (where I pretended to be enchanted with the marvelous headroom and light switches), and I invited him over for a beer at the van. Apart from the mouse, Chuck was my first guest.

"Campus security has got to know about this," he said, sitting on my backseat, laughing.

"You think?"

"Well, that's just my theory," he said. "They're probably just leaving you alone."

"I don't know. I don't think they'd even begin to think someone was living in here."

Chuck and I said good-bye and shook hands. Soon, he'd move up to Worcester, Massachusetts, and I'd be off to Alaska.

"We can go to Walden Pond when you visit," he said.

"Sounds good, Chuck. Take care."

I lit up my stove and started boiling some water so I could cook up a pot of spaghetti stew. It was late April and rather hot, so I opened the windows, shut the blinds, and took off my shirt and pants.

This was the most content I'd been at Duke: I had a friend, I'd gotten through my first semester debt-free, I had a great job tutoring kids, and I had a little money in the bank. Plus, I was going to go back to Alaska and would make a ton of money this summer. School was working: My brain felt sharp and exercised. I came to love my van as much as a person can love a hunk of metal.

Wow, I thought. I might actually miss this thing. I even began entertaining the prospect of living in the van for the rest of my time at Duke. Perhaps I could live in this thing forever?

And that's when a car parked right next to me. Something immediately felt very, very wrong. It was unusual that anyone ever parked directly beside me. Now, especially, because the rest of the lot was empty. I carefully and slowly pulled the blind open so I could see who it was.

It was a white car with blue lights on top. It was a cop. Or a security guard. Whoever it was, I'd been found. I immediately turned my stove off and laid my body flat on the ground so I could stay out of view.

Please, please, please, don't knock on my door! Don't knock on my door!

His car door slammed and he began walking toward the van.

While it was true that my semester was almost over and while it was true that I wouldn't have to live in a van next semester, deep down, I wasn't ready to end my experiment or sell the van. I didn't realize how much the van meant to me until I had reason to worry it could be taken away.

I listened to his footsteps get louder but then, to my relief, softer and softer.

Part IV

VANDWELLER,

or

How I Learned to Live Simply

20

....

RANGER

Summer 2009—Noatak River, Alaska
SAVINGS: $10,000 AND GROWING

THE CESSNA CARAVAN BUZZED noisily though a corridor of treeless, moss-bearded mountains. I craned my neck to look out the passenger-side window to watch the plane's shadow below, a black arrowhead sailing over the sunny green tundra valley floor, wafting over creeks, and turning oblong and abstract when passing over stands of spruce trees.

It was the end of my second summer as a backcountry ranger. In a couple of weeks, I'd have over $15,000 in the bank: more than enough money to pay for next year's tuition. Soon, I'd be out of the van and the wild, and finally inside a cozy apartment of my own.

I was about to be dropped off in the wilderness with Whitney, a Park Service volunteer who'd been helping out at the ranger station in Coldfoot. She and I were to paddle eighty miles of the Noatak River to the Gates' western border, where we were told to keep our eyes out for sheep poachers.

The Noatak River is 420 miles long, stretching from the

glaciers of the Gates' tallest mountain (the craggy, windswept 8,570-foot Mount Igikpak) to the Chukchi Sea on the western edge of Alaska. Because the Noatak National Preserve (6.5 million acres) is connected to the Gates of the Arctic National Park, the Noatak, which runs from one park and into the next, is one of the largest protected watershed basins in the world.

Our pilot gently dropped the plane's aluminum floats down, skimming a placid lake that was a short distance from the Noatak. He helped us unload our gear before he got back into the plane and flew to the village of Bettles, a small native community just outside the park where a ranger station was located and where I'd been living when I was not on patrol.

From the lake, we had to portage our inflatable canoe and gear a mere quarter mile to the river. We had a large inflatable rubber canoe, canoe repair equipment, paddles, communication devices, and a shotgun, not to mention tents, clothes, sleeping bags, and food for the next eight days. I filled up my dry bag with everything I could, slid my arms through the shoulder straps, and marched toward the river, stepping carefully over tussocks and elbowed my way through a six-foot-tall tangle of willow and alder trees.

Whitney was far behind, so I arrived at a grassy clearing next to the river first. When I arrived, I knew I wasn't alone. I felt the presence of something big. Out of the corner of my eye, I saw something enormous and brown, an immovable tree trunk of muscle and fur. At first, I pretended not to see it. Unconsciously, I made the decision to focus first on dropping my eighty-pound pack, which I tried to lift off of my shoulders as casually as possible. It wasn't until I'd shed the pack that I gave the creature – fifteen yards away – all of my attention.

I instantly realized that I was unarmed. I'd left the shotgun back with the rest of the gear by the lake. I knew the canister of bear spray that dangled on my hip was useless because the wind – which was blowing so forcefully that it felt like it might dislodge my NPS ball cap and Frisbee it into the Noatak – would cause the cayenne pepper spray to blow right back into my face and blind me.

I was standing in front of a grizzly bear. And I was completely defenseless.

Its size was flabbergasting. On four legs, its head came as high as my chest. It was ugly: mottled black and brown, with thick, round jaws, a long black muzzle, and two fuzzy semicircles for ears. An unflagging stream of wind combed over its hide, making its fur flutter with aurora-like irregularity. Its hair – millions of little joysticks – twisted up and down, left and right. It was as if the bear was choreographing the movement of each bristle.

It kept its head low to the ground. Its eyes . . .

Whitney suddenly broke through the alder thicket behind me. She paused, quickly discerning that the bear and I were locked in a staring match. The three of us stood perfectly still.

The eye of a grizzly bear. I'd seen this eye before. I'd seen it in the animals, the waters, the mountains. The wild green eye of nature is always spying, always staring, always watching, ready to lock on to yours when you're ready for it. Look into it closely and in the eye's reflection you might catch a glimpse of the objective you: that cultured creature of softness and sophistication or, if you've done well to fall far enough from civilization's grace, the very brute, beastly nature of the wild eye itself. Up until this moment, I'd kept my gaze averted whenever I'd felt that eye fall upon me. But now I was left without an option.

I didn't fear this bear. One doesn't feel fear in such a situation. To call what I was feeling "fear" would do disrespect to what I felt. It was an adrenaline overload, a heightened state of being. Every square centimeter of me screamed with startled life. I was abuzz. Yet, at the same time, oddly composed.

The bear looked composed, too, almost indifferent, as if it had been there and done that a million times before. Yet I knew it must have been feeling something similar to what I was feeling. In a place as wild and remote as the Gates, bears and humans rarely get the chance to interact. This was probably as new and earth shattering for it as it was for me.

It appeared as if it had no intention of fleeing. Why wasn't it running away like the rest of the bears I'd seen up here? Why wasn't I?

It sat down on its meaty haunches and rolled out its tongue like a child's paper streamer. Its teeth were a horrid landscape of stone-gray spires set upon a ridge of slimy purple gums, jutting out of a bubbly white surf of saliva. It was leering at me from its side.

I didn't want to die. I didn't regret a thing, but I didn't want to die. *Give me another day. Please! Just one more!*

I'd walked many miles in the Brooks with this moment in mind. I felt as if this grizzly had been following me, watching me from the woods, waiting for the right time to hold our meeting. For years, the grizzly had lumbered through my dreams. I was never sure why. Maybe it was just plain old fear: knowing that if I walked around in the Brooks long enough I'd eventually have one such encounter. Yet it wasn't simply fear. I revered the grizzly. I spoke of it in hushed tones if I spoke of it at all. Maybe it was because I'd wanted to be everything the grizzly was: wild, strong, free.

I'd wanted to have this experience for years, but only so I could go home and impress people with stories of my foray into the wild and pretend I was some sort of mountain man. But now, in front of this bear, I knew I'd have nothing to brag about. I didn't feel courage or resolve. I only wanted to get away.

I'd come to consider the Brooks my new home, but in front of this grizzly, I knew I was only an outsider, an intruder, an imposter. This wasn't my home. Not the way it was this grizzly's home, at least. As far as I had come, I could see in its eyes that in my life I would never be as wild as it. Yet, after enough time in the wild, I'd feel, if just for a moment, the place begin to overtake me, like vines crawling up the walls of deserted buildings, weeds burgeoning in concrete cracks, whole forests reclaiming the ground that had been seized from them. These were the moments when all the things that made me feel like a stranger, a visitor, a cheechako, would be forgotten as I became more at ease with the land, settling into it like a mammoth tusk buried in permafrost. The savage in me emerged, naked from the woods, full of unchecked lusts and dark desires. Conscious thought would be muted, giving way to deafening instinct. I'd tap into some terrifying source of energy.

I knew then what the heart of man can hold. I felt in me the potential for the best and worst of our nature. I was capable of murder and rape, cruelty and deceit, yet also firmness, assertiveness, courage, determination, vitality. In another lifetime, if I'd been born of the woods, I might have been made coarse and rough and wicked, having been educated in the wild world of amorality, of dog eat dog, of survival of the fittest. But I wasn't this person and never would be. I felt the wild overtake me, but only for moments; my civilized side would always drag me back into the world of right and wrong, discipline and justice. I'd been too civilized to ever be as wild as this place. Yet having become acquainted with the wild man who dwelled inside me, I knew I'd carry him around with me for better or worse, for the rest of my life.

I remembered my training. I called out, "Hey, bear," in respectful tones, slowly waved my arms over my head, and began walking sideways back into the bush, always keeping my face pointed toward the bear.

I saw in my mind what was going to happen. I'd feel the ground shake as it galloped toward me, its paws thudding through the tussocks. I'd look back at the last moment and see its jaws opened wide, its eyes rabid and bloodthirsty.

When we got back to where the rest of the gear was, I loaded the shotgun and waited for it. But it never came.

It let me live.

At that moment, I wanted nothing more than to leave the Brooks and never come back.

• • •

Hours later, we figured the bear had moved on, so Whitney and I hauled the rest of our stuff to the river, inflated the canoe, and began our paddle to the western boundary of the park.

We dipped black plastic paddles into the icy, clear Noatak beneath hundreds of huge, puffy cumulous clouds creeping over the tundra, a flotilla of galleons off to war that – when they crossed the sun's path – flung shadows over mountain ridges with such force it seemed as if they might knock over oblivious hikers (if there were any out there).

Except for the sedge that swayed with the breeze atop the flat-topped bluffs next to the river, the arctic seemed as quiet and empty and lifeless as always. Most of the birds had migrated south, the fish had fled for warmer waters, and the mosquitoes had all died off because the nights had become too cold. But, as always, the arctic would remind me that life abounds: a trio of wolf pups belly flopping into the Noatak; a beer-bellied ground squirrel barking at us from the bank; a red fox staring at us intently, holding a dead vole in its mouth.

To keep our eyes out for sheep poachers, we walked along the hilltops near the Gates' western boundary. Here, the mountains weren't jagged and craggy like they are in other parts of the Brooks. Here, they were smoothly sculpted, round, rolling blobs of waving sedge, squishy moss, and frazzled lichen. We found a good hilltop from which we could look out for potential poachers. But we didn't see any. All we spotted were caribou. Caribou in groups of two, then six, then twenty, then several hundred. Without realizing it, we'd parked ourselves in the middle of the herd's southern migration.

We watched them for hours. Velvet antlers rose regally. Happy families followed one another in dignified processions, nibbling the tundra on the go. A young one, probably just a few months old, came within twenty feet of us, inspecting us with eyes as big and black as eight balls. When we startled a group of three, their tails shot up, a fleeing parade of adorable white asses. Caribou seem to almost defy gravity, trotting, prancing, floating across the country, their hooves barely grazing the ground.

I was envious of the caribou. They spent their lives walking, roaming, eating, exploring. At different points of the year, they changed from solitary travelers to small family units to thousand-strong herds: perhaps the perfect blend of solitude, family, and community. Whitney, skeptical of my whimsies, reminded me of the wolves, the calf mortality rate, the -60°F winters, and the legions of bugs that try to eat the caribou alive, keeping them on the run for most of the summer. I didn't think I'd care about any of that, though. Give me a full life over a long one. I'd give up my retirement years if I could hold up my antlers on mountains like these in my glorious youth.

I looked over the country. We'd reached the western edge of the Brooks. In the distance the hills leveled out to a flat tundra plain, and the Noatak was a canal of liquid gold, set afire by the setting sun, slithering off toward the horizon, lazily wrapping itself around the earth's arc. I thought about how Eskimos and Indians had roamed over arctic valleys like the Noatak for thousands of years, yet there wasn't the slightest sign of nature in disrepair.

I thought of climate change, the unemployment rate, the clusterfuck we call our world. But for just this moment, this day, this week, it was easy to forget about all that. Everything else may have gone to hell, but here in the Noatak Valley the caribou still migrate, the musk oxen still graze, and the grizzly still sits at the top of a healthy food chain. It was the sort of scene that would make one proud of being a member of the human species; we could have built roads and developed this place like anywhere else, but we chose the high-minded and rarely traveled route of restraint. If we can save the Noatak, what other wonders are we capable of?

It had been a week since the grizzly encounter. Whenever I had a particularly scary brush with a wolf or a moose or a bear like I'd just had, I'd feel a need for civilization, for its walls and its security. Yet in time, strangely, it was the memories of these wild encounters that would draw me back to Alaska again and again.

When I thought about my hitchhikes, the voyageur trip, Duke — I was happy to have suffered; I was happy to have been miserable; I was happy to have been alone. And I knew I'd soon be happy to have been scared half to death by that bear. That's because it was in those moments, when I was pushed to my limits, that I was afforded a glimpse of my true nature.

I learned such a glimpse cannot be gotten with half-hearted journeys and soft endeavors. Nor could I hope for such a glimpse merely by setting out to conquer some random geographic feature, like getting to the top of a mountain. Rather, I knew one must confront the very beasts and chasms that haunt our dreams, block our paths, and muffle the voice of the wild man howling in all of us, who calls for you to become *you* — the you who culture cannot shape, the you who is unalterable, uncivilizable, pure. *You.*

Koviashuvik is an Inuit word that means "time and place of joy in the present moment." I'd used to think that the word probably meant something like "nirvana," attained only by the Eskimo version of the bald, saffron-robed man on a mountaintop who's able to achieve a state of unity with everything. Maybe that was the case, but more and more, I began to believe that to live a happy present requires having lived a full past. It requires that we go on our own journey. And if we are so lucky as to reach the end of that tortuous, troubled path, we may be afforded the gleaming vista of self-discovery. This, I thought, was *koviashuvik*.

Days before, in front of the bear, I was reminded that I was as civilized as I was wild, and that I was as drawn to humanity's many marvels as I was to the wilderness's. I was headed back to Duke and civilization again, but I wanted it to be different this time. This time I wanted to bring the wild back with me. To do that, I knew I couldn't seal myself in a house or an apartment or any sort of expensive, luxurious box. If I wanted to stay close to nature and my true needs, I would have to continue to live a bare-bones, simple, uncluttered lifestyle.

While living in the van started off as an experiment, it was clear to me that it was a way of life I'd never truly leave.

21

·······················

PILGRIM

Fall Semester 2009–Duke University
SAVINGS: $13,000

I T WAS AN ORDINARY MONDAY at Westwood College's corporate building in Westminster, Colorado. Just the standard babble of business: the pitter-patter of computer keyboards, the bleating squeals of swivel desk chairs, the jumbled chorus of smooth-voiced salesmen peddling their company's products over the phone.

Josh had been working at Westwood for five months. Westwood admission representatives like Josh – let me remind you – were misleading prospective students about graduation rates, the sort of jobs they'd get, and the enormous cost of the degree. The education itself was substandard, with credits that couldn't transfer over to traditional colleges.

The job was causing Josh's relationship with his girlfriend to fall apart. Random spats about laundry blew up into battles of Homeric proportions. The tone of his e-mails to me had become desperate. He wanted to quit, but he just couldn't bring himself to do it.

The dreary white walls. The bland facial expressions. The

corporate jargon of the office employees. The orderly, sanitized nature of the place. It was as if he'd woken up one day to find himself as one of the background drones in a dystopian novel.

Josh nudged his arm against his computer keyboard and noticed a small piece of paper sticking out. Curious, he picked it up and saw that there was a message written on it.

His heart heaved. Josh had slipped this message under his keyboard months before but had since forgotten about it. He looked at it for less than a second. He didn't have to look at it any longer. He didn't have to read it. He knew exactly what it meant.

He shoved the message back under the keyboard as a last-ditch effort to sweep his conscience under the rug. But it was too late. He sat there thinking for the next thirty minutes.

He took an early lunch and headed to a nearby Target.

Written on that piece of paper was a quote from Hannah Arendt, the writer Josh had admired in college. In her book *Eichmann in Jerusalem* Arendt wrote about Adolf Eichmann, a Nazi leader who organized the murders of millions of Jewish people. Eichmann helped the Nazis not necessarily because he subscribed to their ideology, but because he was a man of little character; he was a bureaucrat who followed orders.

Josh had copied the quote down when he'd thought of one of his supervisors, Mick. Josh didn't think Mick was a bad guy. Mick was perfectly likable: a fair boss and reputedly a good father. But, as Josh saw it, Mick had confused doing well at his job with doing well in life. He was determined to get his admissions representatives to excel, even if that meant the students they signed up would become enslaved by debt. It wasn't that Mick was ignoring the pleadings of his conscience; it was just that he didn't have an actual conscience when it came to such matters.

"The trouble with Eichmann," the Arendt quote read, "was precisely that so many were like him, and that the many were neither perverted nor sadistic, [but] that they were, and still are, terribly and terrifyingly normal."

The quote, Josh realized, wasn't about his supervisor any-more. It was Josh who'd become terrifyingly normal. It was Josh who'd become complacent about the evils he was com-mitting. It was Josh who wasn't thinking enough about the stu-dents he was sentencing to lifetimes of debt.

"I think I'm gonna quit today," he told me over the phone, his words linked to one another by a nervous trill.

"No way! Do it, man!" I yelled.

While he'd dreamed of telling off his boss and inciting some sort of mass exodus among his fellow employees, what actu-ally happened was far less dramatic. He walked into his boss's office beaming with pride and resolve, but he was too polite to make a scene, so he waited forty-five minutes as his boss tried to persuade him to stay. Josh said he'd think about it, left the office, grabbed his things – the Arendt quote included – and told his coworkers he'd see them tomorrow, which wasn't the case at all.

When Josh left Westwood that afternoon, he still had more than $50,000 in debt. He still had rent to pay. He still had gas and insurance bills. And he didn't have another job lined up. But despite his many obligations, for the first time in his life Josh was finally free. He would never again let anything – job or debt or responsibilities – get between him and his conscience. And seeing that quote under his keyboard and feeling in his belly the euphoria of rebellion, he knew he'd never again wonder if his costly education was "worth it." He knew now that even though his education cost him tens of thousands of dollars and years of work, it was worth every goddamned penny.

The euphoria, though, was only temporary. He thought about the students he'd persuaded to enroll at Westwood with remorse. (It turns out that Josh was a horrible salesman, but he had managed to sign up something like three people while he was there.) His old debt had lost its importance, but he now had to figure out a way to pay back this new one.

• • •

Meanwhile, for the first time in my life, I had more money than I knew what to do with. After paying my semester's tuition, I had $13,000.

Upon returning from Alaska, I took a $40 taxicab ride from the Raleigh-Durham airport to the van. (I'd paid $200 to a guy I found on Craigslist to watch it for the summer.) I drove it to Sears to buy a set of white tees and a pair of cargo shorts, and then to Whole Foods where I spent liberally on food: organic yogurt, freshly baked bread, gourmet peanut butter. I went to an outdoors store and bought a $90 backpack, a $225 pair of hiking boots, and a $45 headlamp. Online, I bought a $300 camera.

With each charge to my credit card, I rode the "buyer's high" — a high that I hadn't experienced in years. It occurred just after each purchase, and, like the addict's hit, I felt the gush of instant gratification, followed by a guilty hangover — a hangover only to be cured with yet another purchase and another after that. It was easy to live frugally when I didn't have any money. Being *voluntarily* poor was something else entirely.

I was now a head tutor at the elementary school, still putting in close to twenty hours of work a week, teaching many of the same boys and girls I'd worked with the previous semester. I enrolled in two courses again: one was an undergraduate creative writing course called Travel Writing, and the other was a liberal studies course titled Emotion, Morality, and Human Nature. For my writing course, we wrote stories about our travel experiences and read them aloud in class. I wrote essays about my voyageur and hitchhiking adventures and, for my final paper, a story about living in my van. Afterward, I asked the class to keep my confession a secret, but my professor said, "That was really good, Ken. You should think about publishing it."

Throughout the semester, I dealt with typical vandwelling discomforts: September heat, November cold, loneliness, sexual frustration, squirrels on my roof, and the landscapers who'd

rev leaf blowers around my van every Monday morning, oblivious that someone was trying to sleep inside.

After my last class of the 2009 fall semester, I flew to Boston for
winter break, where I went to visit Chuck, my liberal studies
friend. Together, we drove to Concord so I could finally embark
on a holy pilgrimage to my idol's pond.

It was a mere mile-long walk from the town of Concord,
which, since Thoreau lived there a century and a half before,
had become a bustling warren of cafés, stores, and restaurants,
where shoppers, turbaned under layers of wool and polyes-
ter, entered and exited as if on holy missions of their own. Not
only had some of the businesses borrowed Thoreau's name,
but a picture of his iconic neck-bearded portrait was plastered
on posters, T-shirts, and buttons. (How ironic was my urge
to buy!)

Cars curled around the curves of Walden Street, along which
horses and buggies had clip-clopped and creaked ages ago. At
an intersection, tires plowed through slush, SUVs groaned to a
halt, and engines purred while drivers waited for green.

The trails around the pond, during pleasanter weather, are
typically clogged with fellow pilgrims, but freezing temperatures kept other would-be pilgrims at home, giving us the whole
place to ourselves.

I was surprised to see how big the pond is. At sixty-one
acres, it's actually more like a small lake. Brittle, translucent
ice formed along its perimeter. A covey of ducks flurried to the
pond's liquid center, agitated by the sound of our boots plowing
through rust-colored leaves. The forest, partly made up of long,
slender pines extending bushy green needles, blotched an otherwise bleached-white sky.

If I had been in a dreary mood, I would have called the scene
dreary. But as we edged around the pond, I was excited about
the anticipation of something — exactly what, I wasn't sure. Perhaps I imagined the ghost of Thoreau rounding the next bend of

the trail on his midafternoon walk, silently nodding to us without slackening his gait.

My strongest impression of the place was how close we were to town. Not only that, but evidence of our frenetic culture was everywhere. An Amtrak train rumbled by. Planes screamed overhead. We could hear the hum of traffic from all corners of the forest. Thoreau, back in the 1840s, certainly wasn't bothered by any of this clatter, but it was obvious that he was hardly separated from society.

And that's what struck me most about his experiment. He wasn't living in secret, or far away in the woods. He was surrounded by society. For everyone to see. And while it seemed he was living only for himself out there in his cabin, I now realized that this experiment wasn't just for him. It was for everyone.

I'd been living for myself for years now: paying off my debt, going on adventures, enrolling in college, and saving money for my own endeavors. Yet what I desired now more than anything wasn't more adventure, or excitement, or money. It was purpose. I wondered what the purpose of this past year in the van had been. To save money? To have a zany experience that I might joke about with friends someday? I felt I had learned something in Alaska and in the van, but it seemed so wasteful to keep it all to myself. While I didn't think there was anything wrong with the hermit who kept to himself out in his cabin in the woods, I didn't think I could be satisfied to live a life in which I played no meaningful role in other people's lives. Unlike the hermit, the ascetic who lives in his woodland cabin or wears his homespun cloth – and does so in the midst of civilization – has the important duty of sharing his experiences. So with Thoreau in mind, I decided to publish the article I wrote for my class on the online magazine Salon.com – where my professor knew one of the editors – revealing the secret I'd kept for almost a year.

The day after it was published, I had eighty-six new Facebook friend requests, more than a hundred messages and e-mails, and media outlets calling me for interviews. The story had gone viral over the Internet, and I experienced something close to fifteen minutes of fame. I was an object of adoration: "It was your picture that caught my eye," said one admirer over e-mail. "I wanted to drop you a note to tell you my heart flutters each time I look at your picture." This was from a dude, actually, and all other amorous advances, curiously, came from homosexual men (but compliments taken nonetheless). When I was walking down Ninth Street, one woman from her car cheeringly called out, "Hey, van man!" While interviews with me made the front page of the Buffalo- and Raleigh-area newspapers, I told Fox News to go to hell when they contacted me. And after watching my first episode of *Inside Edition* (90 percent of which was about Tiger Woods's affair), I turned down their offer of money for an exclusive interview.

My predominant thought: *What's the big deal? Why all this fuss?* I knew there was nothing remarkable about living in a van. There were probably a billion people, after all, living in tighter, smellier, less-hygienic dwellings. Plus, for the great majority of human history, our ancestors lived under animal hides, within dirt walls, in freezing caves. I had a metal roof, a bed, and even an engine and tires that could take me wherever I wanted to go. *Hell, I had a TV.* Yet I knew from the many messages I'd received that it wasn't the van that people were drawn to. It was the freedom the van gave me.

The popularity of the article made more sense when I considered the historical context of the times we were in. We were in the middle of the Great Recession. Everyone was in debt. Many were underemployed. Values were changing. Ideologies were shifting. This is what happens in hard times, I thought. In more prosperous times, if we saw someone growing her own food, sewing torn clothing, or living in an austere dwelling – doing things that a "poor person" might do – we might have felt sorry for her. But in hard times, when we find ourselves in positions of dependence – whether on friends or family or the government — we are no longer so quick to associate frugality with poverty. We can see, rather, that it's the frugal who are immune from economic epidemics. Frugality becomes a virtue, not necessarily because we admire frugality, but because we admire independence.

How fickle we can be with freedom! How thoughtlessly do we surrender our autonomy! The fanciness of our dress, the make of our car, the brand of our gadgets, the name of our school. We spend our savings or go into debt for no other reason than to bask in the warm rays of peer approval. Yet fashions are slavishly followed one day and ridiculed the next. Be a devotee in the Church of the Consumer and you'll forever live in fear of the capricious God of Style. Freedom, though, is an honest pair of eyes, a healthy physique, a cheerful laugh. Style goes out of style. Freedom is forever.

Still, though, now that I had money, I became restless. Mostly, I fantasized about van renovations. I thought that it would be nice to have solar panels or a wood stove. It would be nice to have a bike rack, a new paint job, and an Internet connection.

Someone once offered Thoreau a welcome mat, but he declined because he preferred "to wipe [his] feet on the sod before [his] door," adding, "It is best to avoid the beginnings of evil." While I understood Thoreau, in his writings I began to notice an unflinching unreasonableness, a rigid ideology, a foolish dogma. And while he may indeed have rejected that welcome mat, he left out of his book the fact that his mother did his laundry for him. He fashioned himself on the page as an icon of stern independence, but Thoreau, in reality, was just another guy who didn't like to do his laundry.

This made me think that I might have been unreasonable, too. For the past year, I'd refused all offerings of help from my family, doing away with the whole gift-giving, gift-getting ceremonial act, which is an act that has been forging and fortifying human relationships since the dawn of man. And while I really did want to stay out of debt, I acknowledged that accepting a gift or going into debt was an okay thing to do in some circumstances. By borrowing money today, we can invest in a house, a farm, an education, or a business, so we can live happier, hopefully debt-free lives in a more prosperous tomorrow. And while Josh and I despised our debts, the educations we'd bought with the money we borrowed were priceless. A college education is one of the few purchases a person can make that cannot be repossessed or auctioned off.

It was easy for me to see now that when we try to be a "Thoreau" or a "minimalist," or when we live according to a strict ideology, we begin to confuse someone else's needs with our own. So to live in harmony with my own particular needs and desires, I knew I had to test ideologies, not follow them. I told myself that it was okay to want things and, if I had the money, to buy things. But I knew better than to fall victim to "it would be nice."

I knew all about "it would be nice." I saw it everywhere. A middle-class family might think it would be nice to have an in-ground swimming pool. A millionaire might think it would be nice to have a yacht. A billionaire, a private jet. The desires never stop.

I knew that there were people living in real poverty — people who really could use an opportunity to "move up." Someone, somewhere, might think it would be nice to have food to feed her family. Someone, somewhere, might think it would be nice to be enrolled in college. Someone, somewhere, might think it would be nice to have potable water to drink, a job to work at, and a roof over his head. Someone, somewhere — I was sure — might think it would be nice to be in my situation. What if I thought it would be nice to be me, a vandweller? My journeys, the time in Alaska, the year at Duke, Thoreau: from them I learned that I must appreciate what little I have instead of restlessly longing for what I did not.

Even though I no longer thought so ill of gifts and possessions and debts as I once did, I still wanted to see this thing through.

I wanted to get my degree debt-free. Like I'd planned. And even though I had money, I knew that getting through Duke debt-free would still be no simple task. I had another six courses and three semesters to go. Tuition and books would cost me another $7,000 alone.

—1.5 Years Later—

22

....

GRADUATE

May 2011—Duke University
SAVINGS: $1,156

O N MAY 14, 2011, I woke up on the floor in the corner of
an Embassy Suites hotel room in Raleigh, North Carolina.

The night before, my mom, dad, and aunt had flown to North
Carolina from Buffalo. They had planned on sharing a king-
sized bed while I slept alone on the pull-out couch in the other
room. But just minutes after the lights were turned off, I heard
some terrifying snarls that sounded like dinosaurs moaning in
battle from the bedroom.

My aunt, we then learned, had a snoring problem. It was a
roar that, with each passing sound wave, made the paintings on
the wall rattle and my hair flutter.

Justifiably, my dad left the room to join me on the pull-out
couch. But he was—I was afraid to learn—just as bad, emitting
enough sniffles and snorts to make me think I had bedded down
with a pride of slumbering lions.

This time I left the bed, grabbing my pillow, snagging a towel from the bathroom, and sandwiching my head in between them in the farthest corner of the room.

Poorly rested, I woke up anxious and groggy. It was the day of my liberal studies graduation – the conclusion to my two-and-a-half-year vandwelling experiment. In a couple of hours – to my absolute horror – I was to give a speech as my graduating class's student speaker.

After publishing an article about my first two semesters at Duke, part of me – despite my social disinclinations and solitary penchants – hoped to enjoy a period of fame as the campus "van man." I imagined myself walking around campus giving high-fives to strangers, fist-bumping professors, and shaking hands with janitors. I'd walk by a group of girls and they'd cup their hands around their mouths to hide coquettish smiles. Perhaps I'd even take a disillusioned undergrad under my wing. Maybe I could be the campus sage? I'd wear loose white clothing, grow out my beard, and speak in aphorisms to any underclassmen who'd journey the mile on foot to my sacred parking space.

Things, needless to say, didn't turn out this way.

Duke administration, not sure how to handle the situation, told a local newspaper that they were prepared to offer me "guidance and counseling." A tenant in the apartment complex by my parking spot complained to the landlord that my presence made him or her "uncomfortable." I was kicked out of the Mill lot, but I was generously given a new parking spot in the middle of campus (but only if I signed a contract saying I wouldn't sue Duke and that I'd never live in the van on campus again after graduating).[1]

Eh, fair enough.

I lived in the van for another year. I was visited by another mouse, I dealt with a swarm of ants that overtook my storage

---

1   As a result of my experiment, Duke has prohibited students from living in their vehicles on campus for reasons of, as one Duke official explained, "safety, security, health, and liability."

container, and I salvaged food from the Student Union garbage cans whenever I was hungry and low on cash. I took two more creative writing courses and other random, enriching courses like History of Economic Theory and History of Sincerity, and did an independent study on "Student Debt and the Self."

After two years in the van (and my final semester spent living on a small farm in rural North Carolina, where I wrote my final project), I completed my goal of graduating debt-free. I ended my experiment with $1,156 in the bank and about the same number of possessions I had arrived with. I had a monetarily useless degree, no real home, and hardly anything to my name.

Thoreau said he found it hard to leave Walden Pond but decided to move out because he had "several more lives to live." Like Thoreau, I thought I could have happily remained in the van, but I felt that I had more lives to live, too. What those lives might be, I wasn't sure.

But the decision about what life I should live took on an undesired urgency when a respectable magazine, whose editors were acquainted with my Salon article, insisted that I apply for a writing job that paid in the high thirties. Their one stipulation was that I had to make a three-year commitment. Just like that, I could have an apartment, health insurance, a stable year-round job, a decent salary, a comfortable, respectable life.

The graduation ceremony was held at the Washington Duke Inn, a grandiose hotel near campus next to an eighteen-hole, 120-acre, Ireland-green golf course. Men in starched livery opened the doors for my family and me. We walked past bronze busts of Duke presidents, framed pictures of the Duke family, and olive drapes, each set probably worth more than my van. I was used to Duke's extravagant frills, but my parents oohed and aahed admiringly at the ornate chandeliers, the freshly cut flowers in vases, the desks made of bird's-eye maple with a burled walnut veneer.

Family members were dressed in their crisp Sunday best and soon-to-be grads swished around in dark robes. Beneath my robe, I wore my sky-blue dress shirt that I'd bought for $3 from the Salvation Army, the fourteen-year-old pair of dress pants I wore to my high school homecoming dance, dress socks I got for free from the Park Service, and a pair of old beat-up brown shoes. (The origins of my tie and underwear were unknown.)

Fellow graduates and I paraded into a banquet hall, where we were escorted to our seats. Mine was in the front row. As a member of the faculty began the ceremony, I nervously leafed through my speech one last time.

As I sat there, awaiting my turn to speak, I thought about all the people who had played a role in my story. While Thoreau downplayed the society around him, calling them "insignificant," not a day went by when I didn't think of the people who'd been a part of my journey.

Jack Reakoff, the wise man of Wiseman, is still hunting and trapping and giving tours for Coldfoot. James, the seventy-five-year old vehicle-dweller, still lives in Coldfoot and works for the BLM, but he has upgraded from his Chevy Suburban to a cozy BLM cabin. The Gulf Coast Conservation Corps has become defunct for lack of funding, and I've lost contact with nearly everyone I worked with there, just as I have lost contact with my fellow voyageurs and the hundreds of people I hitchhiked with.

Sami took to a life of adventure, canoeing Canadian rivers and biking West Coast roads. Now she works at a ski park and studies at a college in California. She hitchhikes alone, and I tell her that I now know what it's like to go through what my mother went through. Chuck moved to Boston and began working for a nature conservancy. My mom and dad still think I'm sort of insane, but never have they shown the faintest sign of "distancing" themselves from me.

And Josh?

After Josh quit, he became outspoken about the evils he witnessed at Westwood. He got in touch with a law firm that was filing a class-action lawsuit against the school, and they had him star in a few anti-Westwood commercials to recruit other whistleblowers. And in August 2010, when the ruthless sales tactics of the whole for-profit college industry were coming under the scrutiny of Congress, Josh was asked to testify before the U.S. Senate.

Yes, the actual Senate!

He didn't hesitate. He didn't worry about "burning bridges." This would be his grand moment of redemption.

He flew from Denver to Washington, D.C. Waking up in his hotel room on the day of the hearing, he opened his suitcase to learn — to his absolute horror — that he'd forgotten to bring dress pants. With the testimony just an hour away, Josh, wearing a jacket and tie (along with a pair of faded jeans), sprinted through the nation's capital in search of a men's clothing shop. He found a Men's Warehouse and tried to yank the door open. The store had yet to open, though, so he desperately pounded the glass doors. Not having fully caught his breath, he explained his situation to the tailor, bought pants, and then jogged to the Capitol Building, breathing heavily and soaked in sweat.

He sat down at a table in front of the Health, Education, Labor and Pensions Committee, looking up at Senators Tom Harkin, Mike Enzi, Al Franken, and others. His hair was slightly disheveled, and one corner of his shirt collar leaned over to rest on his jacket lapel. Otherwise, he looked all business.

"Now, Mr. Pruyn, welcome to these proceedings," said Senator Harkin.

Josh coughed, thanked the chairman with a quavering voice, and commenced his speech.

"My name is Joshua Pruyn," he said. "I'm a former admissions representative of Westwood College, or, as Senator Mikulski might have referred to it, a 'bounty hunter.'"

He talked about how students had been manipulated and misled, what resulted from the intense atmosphere on the admissions sales floor, and why he decided to quit.

"I quit my job at Westwood on a Monday morning . . . ," he said. "I started admitting things to myself that I'd been avoiding for almost six months. I accepted that I could no longer tell myself that it was possible to work for Westwood and consider myself to be working within any degree of ethical standards. That Monday morning, I walked out of the building and never returned.

"When I left," he continued, "I had no expectation or reasonable prospect for finding another job quickly. I didn't really think about that. I just thought about how naïve I was when I applied for the job – hoping to help students make a better future for themselves through college. Instead, I left fearing the students I had enrolled would end up with a mountain of debt and little or nothing to show for it."

Westwood felt so damaged by Josh's testimony that they denied several of his allegations, called parts of his speech "deliberately false," and asked that his record be removed from the testimony.

Today, Josh is a grant writer for a nonprofit organization in Denver, Colorado, that promotes early childhood education. In six years, he has paid off $55,000 of his debt. We are still best friends and e-mail each other on an almost daily basis. He still owes $11,000.

After the program director's introduction, I was called up to the podium. I rose from my chair and paid careful attention to each step forward, careful not to step on my robe.

I looked at the audience. My hands were trembling, and I'd forgotten how to breathe. I placed the sheaf of papers on the lectern and looked at the suit-and-tie audience. My mom and aunt waved and my dad smiled.

Here I am, I thought, a vandweller among the well-to-do, a high school slacker about to give a speech at one of the greatest universities in the world. Just a few years ago I was falling apart and hearing voices in my head. I was a boring indebted suburbanite. And yet now, Josh and I – a couple of losers in high school – are living fulfilling, inspired, principled lives. Perhaps we'd changed because we were able to leave this world for a bit, this world of jobs and schools and buildings, and got to see a different one: one of mountains and forests and rivers, of the Brooks Range. Maybe it was because we'd somehow managed to bring back the wild with us.

"As some of my fellow graduates may know," I began, "I've spent much of the past two and a half years at Duke living in my van. I'd like to share my story with you, but first, I should address some common misconceptions.

"No – for the millionth time – I do not live in a van 'down by the river.' No, I haven't abducted anyone. And no, my van, I promise you, is certainly not 'a-rockin'" (but please don't come a-knockin' – that would really freak me out).

"I came to Durham in January of 2009, a couple of days before the start of the spring semester. Two months before, I'd finished paying off my $32,000 undergraduate student debt. To pay it off, I worked for almost three years, mostly at low-wage jobs, putting nearly every penny of every paycheck toward my student loans. While I was working, I told myself that I'd do two things whenever I finally paid the debt off. The first was that I would never go into debt again. The second was that I'd enroll in a graduate liberal studies program so I could resume my education.

"So began two different educations. The first was an education in vandwelling, in loneliness, in frugality, in figuring out how to wash my pots and pans without running water. The second was an education in liberal studies, in Diogenes, in Rousseau, in writing, speaking, and thinking. Yet it wasn't long before these educations came together, like two rivers meeting at a confluence and flowing together as one.

"That's the thing about a liberal education: It's a deeply personal affair. And even if we are enrolled in courses that may seem to have little relevance to our lives, the nature of a liberal education – the lectures, the discussions, the writings – tends to bring faraway subjects close to home.

"Some call the liberal arts self-indulgent and impractical without realizing that the classics, the social sciences, the humanities are fertilizers for democracy, and when the arts are scattered onto college campuses, they create a healthy soil into which students can plant themselves and grow into empathetic, introspective, and conscientious citizens.

"Yet when I think of higher education today, I think of a James Joyce quote. Joyce said, 'When the soul of a man is born . . . there are nets flung at it to hold it back from flight. You talk to me of nationality, language, religion. I shall try to fly by those nets.'

"Today it seems there are more nets than ever. Today, students struggle to fly past a poor job market, around unpaid internships, and through the sticky web of student debt that is nearly as wide as the sky itself. And when curricula lack the liberal arts, college itself becomes another net.

"This education has taught me that one does not become free simply by staying out of debt or living cheaply in a large, creepy vehicle; rather, we must first undergo a period of self-examination to see, for the first time, what nets have been holding us back all along.

"Unfortunately, economic realities and political priorities require that most students pay an unreasonable amount of money for their educations even though the great majority of students only wish to better themselves and society. Yet I find it funny — and fitting — that those of my friends who went into nearly inescapable debt to pay for their educations still say they wouldn't go back and change a thing.

"Today, I leave Duke much the same way I came. I have exactly $1,156, no job, and a degree that is — let's face it — not going to have me, or most of us, rolling on a mattress covered in twenty-dollar bills. And to keep out of debt, I've recently put the van up for sale.

"While I am more or less broke, in exchange for the education I have bought, I have received a wealth in return. I speak of the wealth of ideas, of truth — such is a currency without rates, a coinage that will not rust, capital I cannot spend. I may leave this place with empty pockets, but I shall carry this wealth with me whether I am young or old, at home or abroad, housed or homeless, rich or poor, till the end of my days. Thank you."

The crowd cheered and my parents cried. I stepped off the podium and walked back toward my seat. My experiment was over.

In the days that followed, I wondered what I should do. Should I take the magazine job and all that came with it: the salary, the health insurance, the comfortable life? Was it time to finally settle down? I wondered what Thoreau would have done.

He was many things – a surveyor, a naturalist, a handyman, a pencil-maker – but I thought of Thoreau as a writer more than anything else. And his greatest story wasn't one of his essays, or *Walden*. His greatest story, I thought, was his life. He knew that anything is possible when you wield the pen and claim your life as your own.

But the truth is that so few have the privilege to write their own stories. People are born into poverty without hope of redemption. Children are abused and damaged. Disease and war and famine and a million other things prevent them from wielding the pen. But for those of us who can, should it not be our great privilege to live the lives we've imagined? To be who we want to be? To go on our own great journeys and share our experiences with others?

When I'd looked at the austere furnishings in Thoreau's replica cabin – the single bed, the rickety chair, the desk – I thought about how Thoreau had invented this home, this lifestyle, this life! Oh, how many things we can do! Oh, how we can turn the wildest figments of our imagination into something real!

Yet what stories can we write today? My generation was born in a strange age – an age when nearly every blank spot on the map has been explored, when so many of our wild places have been paved over, when there are no honorable wars to fight or frontiers to settle. Our adventures take place in virtual, vicarious video game worlds, or they're tightly crammed into a gap year. Our inner wildness atrophies without a place to exercise it. We are cubicle monkeys and loan drones. Generation Screwed.

Maybe there is no longer a frontier, but for me the frontier is a horizon as wide and endless as it was for the first pioneers. We have real villains who need vanquishing, corrupt institutions that need toppling, and the great American debtors' prison to break out of. We have trains to hop, voyages to embark on, and rides to hitch. And then there's the great American wild – vanishing but still there – ready to impart its wisdom from an Alaskan peak or a patch of grass growing in a crack of a city sidewalk. And no matter how much sprawl and civilization overtake our wilds, we'll always have the boundless wildlands in ourselves to explore.

This graduation was different from the last one. This time, I was debt-free. But I was more than debt-free. With the whole world within my grasp and ready to be seized, why should I call this the end of my adventures when it can just be another beginning?

I turned down the job offer and said good-bye to friends, family, and professors. I said good-bye to Duke and North Carolina. I said good-bye to my beloved Econoline. Might I have an actual job one day? Maybe. Might I come to live in a home that doesn't have wheels? Probably. I had many lives to live still, and I'd get to them, but just not yet.

On a sunny day in early June, I gave away half my stuff, filled up my backpack, and stepped onto a plane that was headed to Alaska.

I was nervous about turning a new page and starting a new chapter, but I knew both by faith and experience that I'd be okay if I lived simply and kept a light load. I knew I'd be okay if I forever thought of myself as a student, whether seated within the walls of the classroom or on foot through the university of the great outdoors. And most of all, I knew I'd be okay if I listened to the oft-unheard voice within – that wild man who whispers into your ear when you most need it and least expect it:

"Go for it."

ACKNOWLEDGMENTS

Because I was well aware of the limitations of living in a van, after graduation – with the hope of securing a stable Internet connection and source of electricity – I set off on a journey to write this book, hopping from one friend's couch or basement or cabin to another. First and foremost, I must thank the Abbott of Acorn Abbey, David Dalton – master cook, gentleman farmer, dear friend, and unpaid editor extraordinaire – for providing me with delicious meals, grand accommodations, and a garden I could work in. Thanks also go to Coldfoot Camp, as well as its generous and supportive managers, Brett Carlson and Chad Conklin, for letting me be the camp's first-ever "writer in residence." I am grateful for Chuck Johnston for his apartment in Boston, Amelia Larsen for her basement in Denver, Professors Christina Askounis and Bob Bliwise for encouragement and advice, Josh Pruyn for his review (and for letting me use him as a character in the book), as well as Sarah Rice for her companionship and review. I am blessed to have the brilliant JanaLee Cherneski as a friend and reviewer, as well as Peter and Amy Bernstein as agents, who took my farfetched dream and turned it into a reality. I also wish to thank editor David Moldawer for taking a chance on a young wannabe writer who had only a couple of published articles to his name, and copyeditor Nancy Tan for an invaluable last-minute edit.